BASIC Computing
for civil engineers

BASIC Computing
for civil engineers

W. M. Jenkins
Division of Civil Engineering
The Hatfield Polytechnic

J. M. Coulthard
Department of Civil & Structural Engineering
Teesside Polytechnic

G. C. de Jesus
Department of Civil & Structural Engineering
Teesside Polytechnic

VNR (UK) Van Nostrand Reinhold (UK) Co. Ltd.

Published by Van Nostrand Reinhold (UK) Co. Ltd.
Molly Millars Lane, Wokingham, Berkshire, England

Library of Congress Cataloging in Publication Data

Jenkins, W. M. (William McLaren), 1927–
Basic computing for civil engineers.

Includes bibliographical references and index.
1. Civil engineering—Data processing. 2. Basic (Computer program language) 3. Microcomputers. I. Coulthard, James Mark. II. Jesus, Giovani C. de. III. Title.
TA345.J46 1983 624′.028′542 83-3584
ISBN 0-442-30558-3
ISBN 0-442-30559-1 (pbk.)

Printed and bound in Hong Kong

Preface

The development of the microcomputer has been at such a rate as to be almost 'explosive' in character. Over a comparatively short period of time, capacity and speed of operation of 'low cost' computers have increased at a phenomenal rate accompanied by an almost equally phenomenal reduction in cost. The personal microcomputer is now available at little more, in relative terms, than the cost of the slide rule of the 1950s or the desk calculator of the 1960s. The enormous computing power and speed of these machines have opened up new avenues of development in all areas of civil engineering. It is clear that the microcomputer is already having a profound influence on civil engineering education and practice. It is necessary to undertake a fundamental review of curricula and syllabuses in civil engineering to take advantage of the impact of the microcomputer.

Most students of civil engineering are taught some computer programming during their course. The programming language usually adopted for introductory studies is BASIC, followed by FORTRAN for more advanced work. BASIC is capable of meeting the needs of most everyday engineering situations and provides a good basis for a much improved organization of everyday computations, particularly in design situations where improved speed and storage capacity can lead to faster, more economical and more competitive designs.

BASIC was developed in the 1960s as an interactive system for small computers and it is now the most popular language for the 'personal' microcomputer and indeed for all introductory studies in computing. The interactive style of computing provided by BASIC is eminently suitable for engineering computations particularly those involved in design. The 'hands on' aspect is also important for by this method, the user rapidly gains confidence and a feeling of being 'in control'.

The logical approach demanded by the computer compels the engineer to 're-think' his approach to the engineering design process. The result of this is to transfer the mundane computations to the computer and free the designer to act at a 'higher level' of decision making in the design process. This radical change in approach requires the development of new attitudes and expertise. The purpose of this book is to help the student to develop his knowledge of computing and computers alongside, and integrated with, his civil engineering studies.

The first three chapters of the book cover all aspects of computers and programming needed by the civil engineer using BASIC on microcomputers. Illustrative programs are included so that the reader can make immediate, 'hands-on', contact with the computer. It is important to gain this experience as quickly as possible and before turning to more serious engineering applications.

Chapter 4 is devoted to computing with arrays, of fundamental importance in most practical applications where large quantities of data are to be stored and processed. Chapter 5 describes a selection of numerical methods of relevance to civil engineering. The selection, though not comprehensive, is representative of the more important numerical methods. The various forms of outputting information from the computer are discussed in Chapter 6 where some attention is given to the subject of screen graphics, a rapidly developing and useful facility in engineering applications.

Five chapters of the book are devoted to applications of BASIC programming in structural analysis, structural design, surveying, hydraulics and geotechnics. The applications are chosen so as to introduce the reader to the types of computing situations which exist in each of these areas. All the applications are illustrated either with complete programs or program segments which the reader can build in to his own programs. In structural analysis, the emphasis is on the 'stiffness' method in keeping with the modern trend and in recognition of the power of this method when used on computers. The approach to civil engineering design is represented by structural design where some consideration is given to the fundamental, iterative approach to design. The typical design routine is illustrated by programs for reinforced concrete beam design and pitched roof steel frame design.

Chapter 9 is concerned with surveying and includes illustrations of computer programs for closed traverses and for earthwork calculations. Chapter 10 introduces the reader to computing in the hydraulics field through two examples; one concerned with open channel profiles in gradually varied flow and the second with flow in pipe networks. Computer applications in geotechnics are considered in Chapter 11. This is one of the longer chapters and contains a fairly comprehensive review of the subject with a number of useful programs. The usefulness of the microcomputer in the laboratory situation is becoming well known and important developments are taking place. This is recognized in Chapter 12 where attention is given to the interfacing of the computer with experimental equipment in the analysis of laboratory data.

The book contains a liberal supply of programs both for illustrative purposes in programming techniques and also for use by the reader in civil engineering applications. Some of the programs are deliberately kept simple and unencumbered with the refinements which would be desirable in a more usable version. This is intended to help the reader to find his way through the programs easily and quickly. The building in of 'user friendly' and checking devices can then be an exercise for the reader.

Although some attention has been paid to the differences in the versions of BASIC, most of the programs have been written in PET BASIC with some in APPLE or TRS 80/VIDEO GENIE BASIC. The differences are not great but are important and the programming manuals should be consulted in transferring from one version of BASIC to another. The programs, with the exception of very short program segments, are numbered using chapter number and program number within the chapter. Some of the numbered programs are in fact program segments and would need to be augmented before running on the computer.

It is hoped that the book will be useful to all students of civil engineering who wish to become familiar with the power and versatility of the microcomputer and who wish to integrate this into their civil engineering studies.

Our sincere thanks are due to the many friends and colleagues who have assisted us in the preparation of the material for this book, also to our families who have suffered more than a little neglect during the period of writing it.

W. M. JENKINS
J. M. COULTHARD
G. C. DE JESUS

October 1982

Contents

1

Introduction to Computers

1.1 Introduction

Computers in a variety of forms are now penetrating both working and home environments. Their influence is probably most evident in small to medium sized businesses. Due to the reducing cost of electronic components, these firms are now finding that extensive computing power is well within their budget capability and many are taking advantage of the situation. These computers allow calculations and designs to be checked more rapidly and more accurately than ever before and in certain areas of work can almost eliminate the need for paper. In civil engineering the computer allows an engineer to spend more time examining a problem and less time on the calculations in order to produce the best solution. Many engineers are involved in management in which cash flow analysis, estimating and stock control play a major part in decision making. A substantial portion of this work can also be carried out using computer programs.

Engineering problems are often unique and require specific programs to be written in order to obtain a solution, although there are some programs which are written to solve a variety of general problems. In the field of management, existing programs can be rapidly implemented and used. Thus an engineer should not only be capable of using existing programs but also be proficient at writing programs.

The object of this text is to introduce the reader to the programming language BASIC and to develop programs and programming techniques in several areas of civil engineering. BASIC is one of many computer languages and although it is relatively simple to understand, it is also a powerful and versatile language which can be used for both scientific calculations and business data-processing applications. As with any other computer language it does have its disadvantages but one of the reasons for its popularity is its availability on the new generation of desk top microcomputers.

1.2 Computers

Superficially, computers appear to be complex machines capable of intelligent decision making and highly developed understanding. In reality the fundamental processes which take place *inside* the computer are extremely simple; principally basic arithmetic and logical decisions based on simple comparisons. Computers can store large quantities of information, mainly numerical but also alphabetical, can process such information through arithmetic and logical operations and produce the results in intelligible form. Although the basic operations are simple, the set of instructions which the computer must obey can represent a highly complex operation.

Computers should be visualized as machines capable of storing large amounts of data which can be used to produce some type of output, either on a visual display unit (VDU) or paper. Thus to solve a problem a clear set of instructions and data must be

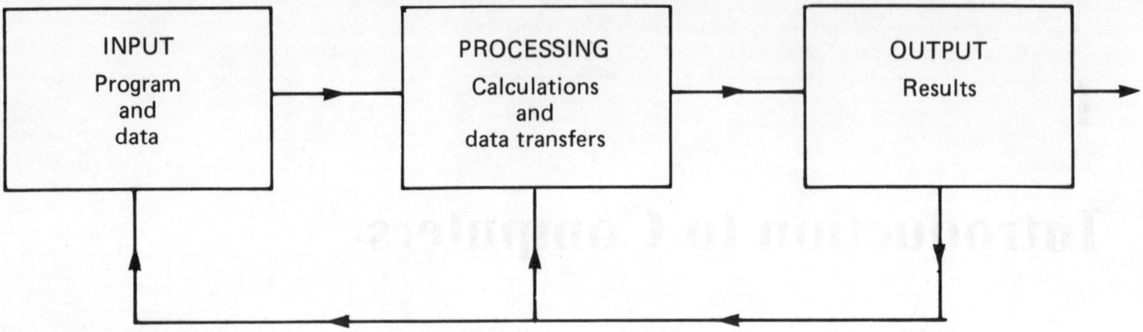

Fig. 1.1 Computer processing cycle.

given to the computer. The set of instructions is called a 'program' and is written by a 'programmer'. The data consist of numbers, words or even electrical impulses from test equipment converted to numbers. The program directs the computer to perform various tasks in a predetermined sequence. The data entered are called 'input', and 'processing' is the manipulation of the data to produce results. The communication of the results to a user is known as 'output'. These concepts are illustrated in Fig. 1.1. Frequently the output from the computations is used to initiate fresh calculations or to modify the calculation procedure. This 'feedback' principle is shown in Fig. 1.1.

At first sight, computers appear to be extremely clever but in reality they can do nothing unless told to do so. Their superiority over the human brain is the speed at which they can perform a given task. The human brain can detect errors in calculations and logic; a computer can only accomplish this if it has been programmed to do so. Thus a computer will produce results that are only as accurate as the program and the data that formed the input, both of which are dependent upon the human.

1.3 Development of Computers

The development of computers has been punctuated by rapid advances in specific areas of technology allowing computers to become cheaper and so increasing the areas of application. Early computers were large and development was concentrated on improving the power of these machines. Such computers were so expensive that only large companies and institutions could afford them. These types of machine are often referred to as 'mainframe' computers. Their operation requires trained personnel and so for the majority of people computers are machines around which a mystique has developed. This mystique was encouraged by the jargon that users and operators built up, and continue to build up.

In the early 1960s the advent of the minicomputer started a trend towards miniaturization enabling smaller firms and laboratories to utilize the computing power that previously had been denied them. Once the development of large-scale integrated circuits had been initiated it was soon possible to mass produce smaller and more complex circuits at a relatively cheap cost. The microprocessor is the continuation of the trend towards smaller computers. Basically it is a device that has all the functions and controlling ability of a large computer on one or several silicon chips. The microprocessor must be connected to additional devices, such as a VDU, to form a microcomputer. The microcomputer and any additional electronic devices are referred to as 'hardware'. The low cost of these microcomputers and associated hardware now makes computing power widely available.

1.4 Computer Configuration

The configuration of the hardware constituting a compter can be visualized as a number of individual units linked together as shown in Fig. 1.2. In fact all computers will have the same basic configuration. The main difference between a mainframe computer and a microcomputer being the size, speed and mode of operation.

1.4.1 Central Processor Unit

The 'brain' of a computer is the central processor unit (CPU). This is the unit that carries out all the calculations, but it also contains the control unit that arranges the electronic switching that will allow data to be transferred between itself and any other device.

1.4.2 Input Devices

These can take many forms and are able to transmit and, in some cases, receive data. Devices such as a VDU with keyboard or a teletype can be used to enter programs and data. They are also capable of receiving and displaying instructions from the computer itself and, therefore, act as output devices. Units which monitor physical phenomena (temperature, pressure, etc.) normally only transmit data when requested by the computer. Punched card readers are again a one-way device but their use is now declining.

1.4.3 Memory Unit

The memory unit is the storage area where programs and data can be stored prior to processing. This unit can be visualized as a series of boxes, each box or 'storage location' having a number which is known as an 'address'. These addresses are used by the operating system and this is a program stored in a reserved area of memory enabling the CPU to manage the computer system. The computer, therefore, knows the memory address at which a program can be found and where all the required data for that particular program are located.

The memory unit is also referred to as 'immediate access' memory as any program or data can be accessed directly by the computer.

1.4.4 Backing Store

Backing store is essentially an extension of the memory unit. It contains data or programs which can be loaded into the main memory unit to be used by the CPU and the operating system. As the CPU cannot access the backing store directly, any information must always be loaded into the main memory unit. Backing store is some form of magnetic medium such as magnetic tape or disks of varying size and performance capability.

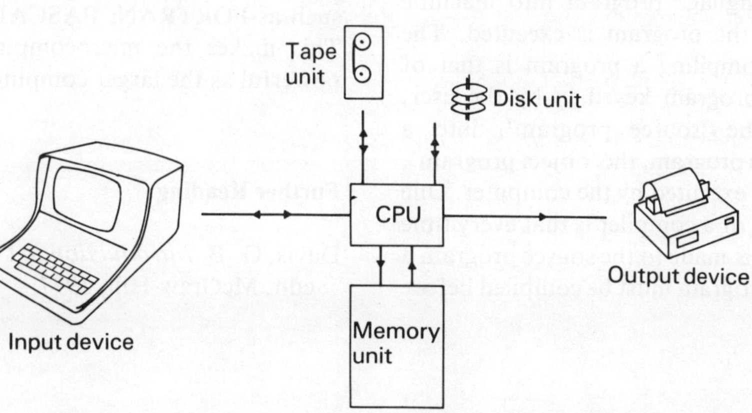

Fig. 1.2 Computer configuration.

1.4.5 Output Devices

Results must be transmitted to the outside world via an output device. This can be some form of printer (teletype or VDU) or a graph plotter.

1.5 Programs and Program Execution

The CPU can only obey a specific set of simple instructions. The instruction set is referred to as 'machine code' and the CPU can execute them at very high speed. In order for the computer to perform complex operations, it must be given a series of these instructions. Unfortunately machine code is almost incomprehensible to the average person and computer languages have been developed which are rather closer to English and thus easier for the programmer. These languages are called 'high level' languages, for example FORTRAN, ALGOL, PASCAL, BASIC. The decision as to which language one should use depends on the problems to be solved, as each language has its own special features. For the reasons outlined in Section 1.1. this text will concentrate on the BASIC language.

A program written in a high level language is incomprehensible to the computer and there are two types of special programs which allow the CPU to run high level language programs.

The first is the *compiler* which translates a high level language program into machine code before the program is executed. The process of compiling a program is that of turning the program keyed in by the user, known as the 'source program', into a machine code program, the 'object program', which is then executed by the computer. One disadvantage of a compiler is that every time an alteration is made to the source program a new object program must be compiled before execution. An advantage, however, is the speed at which the object program can be executed.

The other special program is an *interpreter*. An interpreter is a program that reads and interprets the source program line-by-line. Once a line is read and interpreted the various operations are then carried out. The advantage of this type of program is that alteration and execution of programs can be made simple. The disadvantage is the slow speed of program execution compared to a compiled version. The reason for this is that the interpreter has to spend a lot of time scanning the source program, searching for the data it requires and deciding on the actions it must take.

The high level language BASIC was initially designed for the teaching environment and speed was not an essential prerequisite. Computers supporting several programming languages will have the compiler programs read into a special reserved area of the memory unit. Microcomputers generally handle BASIC by an interpreter. In these machines the interpreter is stored in an area of memory which cannot be erased even when the power is switched off. This kind of program, permanently embedded in a silicon chip, is called 'firmware', as apposed to 'software' which is a program which must be read into memory. Although interpreter BASIC is the standard language for microcomputers it is possible to obtain compilers for languages such as FORTRAN, PASCAL and BASIC. This makes the microcomputer almost as powerful as the larger computers.

Further Reading

Davis, G. B. *Introduction to Computers*, 3rd edn. McGraw-Hill, 1977.

2

Elements of BASIC

2.1 Introduction

The BASIC computer language was developed in 1963 by Professors Kemeny and Kurtz at Dartmouth College USA. In the early 1960s the available high level computer languages were complex and their task was to develop a language that was easy to learn and use. Due to its origins it was given the title 'Beginners Allpurpose Symbolic Instruction Code', hence BASIC. The language initially consisted of only 20 keywords and the common mathematical symbols but the interest attracted from educational establishments and business firms compelled the originators to expand the initial implementation of the language. At the present time almost every large computer facility is equipped with a BASIC compiler or interpreter. Its popularity and use has been further extended by the electronic industry which adopted this language as the standard one supplied as firmware within the new generation of microcomputers.

In microcomputers the main version is that implemented by Microsoft, however, just as the English language has many dialects so there are numerous variations of BASIC. The number of different dialects or versions of BASIC could make learning the language difficult, but thankfully there is a set of instructions common to all versions. The standard set of instructions and keywords will be described in this chapter and the common extensions to BASIC found on microcomputers will be briefly described in Chapter 3.

Once a student has learnt the basic set of instructions and keywords he should then consult the BASIC manuals of the version implemented on the computer he is using. Only by doing this will a full understanding be gained. Having mastered the language on one machine, a user should have little difficulty in transferring to another version. One text that can facilitate a painless transfer is *The BASIC Handbook* by D. A. Lien. This text discusses in detail every instruction used in BASIC thus enabling programs to be converted to run on any BASIC system.

The most important feature of BASIC which makes it popular is that it was designed as an *interactive* language. This allows beginners to write programs and run them quickly and simply, permitting mistakes to be spotted and corrected rapidly. Thus, the learning stage can be accomplished more quickly. Another advantage is the similarity of BASIC to the FORTRAN computer language. FORTRAN is also used extensively by engineers and a knowledge of BASIC can make learning FORTRAN much easier.

A complaint sometimes advanced by professional programmers is that BASIC leads to badly constructed programs. This is true, but with care and a disciplined approach this can be minimized.

2.2 Elements of BASIC

BASIC consists of a series of keywords and common mathematical symbols that must follow a set of simple but precise rules (or syntax). The keywords can be subdivided into those used as direct commands to the com-

Program 2.1

```
10 LET X=5
20 LET Y=8
30 LET Z=X+Y
40 PRINT Z
50 END
```

puter and those used in statements within a program. A program written in BASIC consists of a series of instructions or statements made up of keywords and symbols with each statement prefixed by a line number. Program 2.1 is a very simple program to add two numbers and display the result and this exemplifies the fundamental principle.

Each line of the program must be entered from a keyboard. When control is passed to the BASIC interpreter, the number at the beginning of the line is automatically translated as a line number and the whole line is then stored in the computer's memory as a line of a program.

The program can only be run or executed if the command keyword RUN is entered. A mistake often made by beginners is to expect that the computer knows exactly what is being typed. One must remember that control must be passed back to the computer after entering information from the keyboard. This is done by pressing the 'Return' key, which may also be labelled 'Newline'. If a line number is omitted then the interpreter will attempt to carry out the instruction immediately hence command keywords should not be preceded by a line number.

We will return to 'statements', 'line numbers' and 'commands' again but first let us examine some of the elements of the BASIC language.

2.3 Numbers

The majority of versions of BASIC allow two kinds of numbers; *integers*, which are numbers without any fractional part and *real* numbers which can have a fractional part. Real numbers are also commonly referred to as 'floating point' numbers. Very large or very small numbers are often represented in displays using scientific notation. The conversion to scientific notation depends on the capability of the computer, however, with the modern microcomputer the display handling routines within the BASIC implementation normally handle nine digits. Thus, the number 0.0000023654 would be converted to 2.3654E-06.

The actual size of a number that a computer can handle also varies and one should consult the instruction manuals. In most cases the maximum and minimum numbers permitted should not pose any problems. The limits for two common microcomputers are shown in Table 2.1.

The limitation on the number of digits displayed by the computer can lead to round-off errors. Microcomputers which handle nine digits in the print routines normally round up when the tenth digit is five or more but there are anomalies and again it is wise to check the user manuals.

2.4 Variables

In Program 2.1 the lines 10 and 20 assign a number to a letter of the alphabet. These letters are known as *variables*. Variables take whatever value they have been given during the execution of a program and can represent either numeric data or text.

Table 2.1 Size Limits of Numbers

	Real number		Integer	
	Largest	Smallest	Largest	Smallest
Apple II	1.00E + 38	– 1.00E – 38	+ 32767	– 32767
CBM	1.70141183E + 38	– 2.93873588E – 39	+ 32767	– 32767

A variable name can have one, two or three characters; the third distinguishing between the *types* of variable. The first character must be a letter from *A* to *Z*; the second is optional but can be a letter or a digit. The third character if present can be either a % or a $ sign. The % character indicates that the variable is an integer and the $ signifies a *text* or *string* variable. If neither the % or $ sign is present then it is a *real* variable. Examples of valid variable names are given in Table 2.2.

Table 2.2 Valid Variable Names

Real numeric	Integer numeric	Text (or string)
A	*A*%	*A*$
D	*D*%	*D*$
AZ	*AZ*%	*AZ*$
*C*1	*C*1%	*C*1$
*Z*5	*Z*5%	*Z*5$

String variables or text variables store alphanumeric characters.

Generally, two characters are used for BASIC names but some implementations of BASIC allow variables to have more than two.

2.5 Mathematical Symbols and Expressions

Variables and constants can be combined by mathematical symbols to form expressions. The standard mathematical symbols are supported on all versions of BASIC and are given in Table 2.3.

The symbol for the exponential sign is the only one that tends to vary from machine to machine. The possible alternatives for the exponential operator are given in Table 2.4.

Table 2.3 Mathematical Symbols (or Operators)

Addition	+
Subtraction	−
Multiplication	*
Division	/
Exponentiation	↑

Table 2.4 Alternative Exponential Symbols

Symbol
↑
∧
∩
**

When variables are combined into an expression there can be more than one operator present. One must consider in what order the expression is evaluated. Consider the following example:

10/2 − 1

There are two possible answers to this calculation 4 and 10. To a computer there is only one, 4. The standard order of precedence is first exponentiation, followed by division, multiplication, addition and finally subtraction. The order of precedence can be overridden through the use of *parentheses*. Any operation within parentheses is performed first and if more than one set is present they are evaluated from left to right. The order of precedence can be shown in the expression:

(4**A* + *B* ↑ 2)/(*C* − *D*)

This expression is evaluated in the following order:

(a) In the first pair of parentheses, first *B* ↑ 2 is calculated, then 4**A* and the results of the two separate operations added together. This gives the value of the expression within the first pair of parentheses and this would be stored temporarily.

(b) The expression within the second set of parenthesis, *C* − *D*, is evaluated.

(c) The result from (a) is divided by that from (b).

2.6 Assignment Statements

In Program 2.1, the lines 1Ø, 2Ø and 3Ø are called assignment statements. In its simplest form a variable is assigned a value as in lines 1Ø and 2Ø. A variable can also be assigned a value after an expression is evaluated as in line 3Ø. In the original implementation of BASIC, the interpreter had to find out what was to be done by identifying the keyword that always followed the line number. For this reason LET was used to signify an assignment statement. In the recent versions of BASIC, LET is optional and thus on some computers *both* of the following statements are acceptable:

```
6Ø LET A = C↑2 + D*E
```

or

```
6Ø A = C↑2 + D*E
```

An important point to note is that an expression can only exist to the *right* of the equals sign as only *one* variable can be assigned a value. If an expression occurs to the left of the equals sign the computer will signal a syntax error.

2.7 Mathematical Functions

In addition to the mathematical operators, certain mathematical functions are also provided to allow more complex expressions in assignment statements. Functions take the form:

Function (*X*)

The value *X* inside the brackets is known as the *argument*, and its value is used by the function routine. Thus, if we assign a value of 4 to the variable *X* and then obtain the square root of this value the program statements would be:

```
1Ø LET X = 4
2Ø LET R = SQR (X)
```

The square root of 4 would be assigned to the variable *R* and this could then be displayed. A list of the more common mathematical functions available follows:

SQR (*X*) Computes the square root of the argument (*X*).

SIN (*X*) Returns the sine of the argument which is treated in radians, thus SIN (30) will give the sine of 30 radians. To obtain the sine of 30 degrees, the degrees must be converted to radians, for example

SIN (30*3.142/180)

Some computers have the symbol π which can be used directly

SIN (30*π/180)

COS (*X*) Returns the cosine of the argument (*X*) which is treated in radians.

TAN (*X*) Returns the tangent of the argument which is treated in radians.

LOG (*X*) Calculates the natural logarithm of the argument. To obtain logarithms to the base 10, then the natural logarithm is divided by the natural logarithm of 10, for example,

LOG (*X*)/LOG (10)

EXP (*X*) Raises the natural logarithm base e to the power of the argument, e^x.

ATN (*X*) Computes the arctangent of the argument in radians. Thus ATN(1) returns the value of 0.785398; whereas ATN(1)*180/π returns 45 degrees.

ABS (*X*) Returns the absolute value of an argument. Hence, ABS (−5.6), gives 5.6.

SGN (*X*) This returns the sign of the argument; thus,
If *X* is positive then +1 is returned.
If *X* = 0 then 0 is returned.

If *X* is negative then −1 is returned.

INT (*X*) This converts a real number to its integer value by truncation thus, 12.367 will be truncated to 12. In some versions of BASIC, INT(*X*) will return the highest integer less than or equal to *X*. In this case INT (5.6) will return a value of 5 and INT (−5.6) will return −6.

RND (*X*) This is the random function which generates a random number between zero and one, using the argument *X* as a seed. The seed is used as a starting point for the random number generator. The random function will vary from one BASIC to another. For example, some versions will return values between limits that are previously set; others do not require an argument as a seed. A user should consult the BASIC manual for the computer being used. The function has applications in statistical problems, for example, the selection of sites for soil sampling and testing when controlling compaction on a large earthwork project.

2.8 Program Input

Computer programs usually require data. These can be either input from a keyboard or already part of the program itself. In the original version of BASIC the only way to put data into a program was to have them stored as part of the program in what is known as a DATA statement. Associated with the DATA statement is the READ statement. Program 2.2 shows the initial part of a program to calculate the cross-sectional area of an I-beam. Line 4Ø calculates the cross-sectional area of the beam and assigns it to the variable *AR*. The variables *A*, *B*, *C* and *D* are the beam dimensions as shown in Fig. 2.1.

Program 2.2

```
10 DATA 150,200,20,20
20 READ A,B
30 READ C,D
40 LET AR=(A*B)-(A-D)*(B-2*C)
```

When the program is executed the first statement on line 1Ø informs the computer that what follows after the keyword DATA are data for use by the program. It then passes on to the next line which contains a READ statement which tells the computer to obtain the values for the variables *A* and *B* from the list in the DATA statement. How does the computer know which of the values to take? In fact it does not; it automatically takes the first two values and assigns them to the variables in the order they appear in the READ statement. Thus, *A* is assigned a value of 15Ø and *B* a value of 2ØØ. Having done this the computer keeps a record of what it has used by an internal pointer that points to the third value in the DATA statement, i.e. 2Ø. When the next READ statement is executed the computer assigns the value of 2Ø to *C* and 2Ø to *D*. It is up to the programmer to put the values in the *correct order* within a DATA statement. If there are insufficient values within a DATA statement a program will be stopped and the computer would signal an error. DATA statements can also contain text

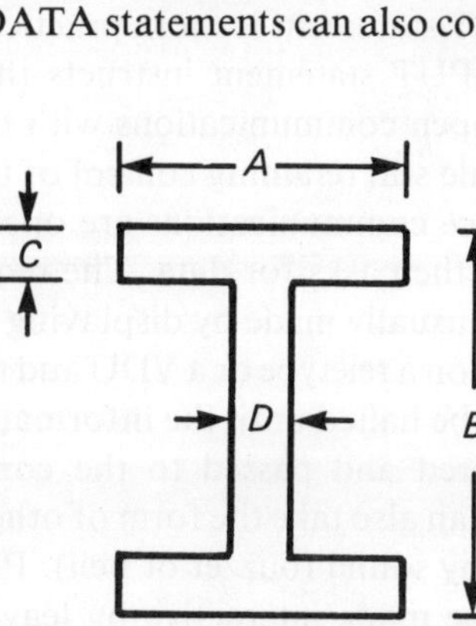

Fig. 2.1 I-beam cross-section.

Program 2.3

```
10 DATA  1979,1980,1981,1982,"YEAR"
20 READ Y1,Y2,Y3,Y4,Y$
30 PRINT
40 PRINT Y$,Y1
50 PRINT Y$,Y2
60 PRINT Y$,Y3
70 PRINT Y$,Y4
80 END
```

variables or a mixture of numeric values and text, as shown in Program 2.3. Any text in a DATA statement should be within quotation marks, although with some versions of BASIC they are unnecessary.

Associated with the READ and DATA statements is the RESTORE statement. As data are read by a READ statement so the pointer progresses down the list of data. If a RESTORE statement is included within a program, then, when it is executed, it causes the pointer to jump back to the start of the data list thus restoring *all* the data that had previously been read. In this way the same data can be read again and again by different READ statements.

The DATA, READ and RESTORE statements are especially useful when testing the execution and output from a program as the data need only to be entered once to become part of the program. For an interactive program where data change from run to run then the INPUT statement is more practical.

The INPUT statement instructs the computer to open communications with the keyboard while still retaining control of the program. Once communications are opened the computer then asks for data. The request or prompt is usually made by displaying a question mark on a teletype or a VDU and the program will be halted until the information has been entered and passed to the computer. Prompts can also take the form of other characters or by sound (buzzer or bell). Program 2.2. can be made interactive by leaving out line 1Ø and replacing the READ statement with an INPUT statement, see Program 2.4.

Program 2.4

```
20 INPUT A,B
30 INPUT C,D
40 LET AR=(A*B)-(A-D)*(B-2*C)
```

INPUT statements are very precise and the type of data entered must be consistent with that requested. An attempt to enter text when a number was expected will cause a response from the computer. The type of response will vary but most will ask for it to be retyped and entered again.

As with READ statements both text and numbers can be entered on the same INPUT statement, however, some computers will require the text to be in quotation marks others do not.

2.9 Program Output

Programs must be able to display information on a VDU or a teletype and this is accomplished by the PRINT statement.

2.9.1 PRINT Statement

The PRINT statement instructs the computer to do just that on its output devices. It can be used to print numeric or text variables or just text. Associated with the PRINT statement are the punctuation characters, commas and semicolons, that are used in arranging or formatting the display. Let us first consider the printing of numeric variables using the following statement:

PRINT X,Y,Z

Assuming that the variables X,Y and Z have previously been assigned values, then, when they are printed they will be spaced out across a line of output. The reason for this is that BASIC divides a line across a page into a series of zones, each zone having a set number of character positions. The number of zones varies from one version of BASIC to another, some have eight zones with ten characters per zone, others have four or five zones. The variables are displayed at preset positions and, providing the values of the variables have fewer characters than the allotted zones, there will be clear spaces between the values. In some instances a large number of spaces between variables may be unnecessary, this is often the case with text and numeric variables printed adjacent. Spaces can be eliminated by separating the variables in a PRINT statement with a semicolon, for example

PRINT X; Y; Z

The effect of the semicolon is to prevent the spaces being inserted and it can also prevent a carriage return and line feed instruction being carried out. In the previous examples of PRINT statements these instructions would have been automatically carried out by the computer. The effect is that the next time a PRINT instruction is encountered the information will be displayed on a new line. A semicolon at the end of a PRINT statement would prevent this so enabling information to be printed on the same line by two different PRINT statements. An example of this can be seen in Program 2.5, lines 14Ø and 15Ø.

The style of printing numeric variables depends on the size of the number, and the number of characters the computer's PRINT routine can handle. Generally speaking numbers close to zero (< 0.01) and larger than 10^6 are printed in scientific notation. The actual values will vary from machine to machine.

The programs shown so far contain just simple statements but the PRINT statement can make them more understandable by printing text and blank lines. A statement containing just the keyword PRINT will produce a blank line of output. Text in the form of instructions and titles can also be printed but this must be enclosed in quotation marks. Various uses of the PRINT statement are shown in Program 2.5 which is an improved version of Program 2.4.

Program 2.5

```
10 PRINT
20 PRINT "CALCULATION OF CROSS-SECTIONAL"
30 PRINT "AREA OF AN I-BEAM"
40 PRINT
50 PRINT "ENTER WIDTH OF BEAM"
60 INPUT A
70 PRINT "ENTER HEIGHT OF BEAM"
80 INPUT B
90 PRINT "ENTER FLANGE AND WEB THICKNESS"
100 INPUT C,D
110 PRINT A,B,C,D
120 PRINT
130 LET AR=(A*B)-(A-D)*(B-2*C)
140 PRINT "CROSS SECTIONAL AREA = ";
150 PRINT AR
160 END
```

2.9.2 Formatting Functions

Arranging or 'formatting' data and headings on a screen or teletype would be rather difficult if only the PRINT statement were available. Fortunately there are other functions that allow us to arrange the displayed text.

The most used function is the TAB function which in some ways is similar to typewriter tabulating and takes the form

TAB (*n*)

where *n* is the column position. The function allows data and text to be displayed at set positions thus allowing columns of data to be aligned neatly on a screen or line printer. To display the information in columns one must first determine the character position on a line where each column is to begin. The first character position on the left-hand side of a screen or paper is usually referred to as position Ø, and the tab position is always relative to the left-hand side of the screen or paper. The following statement will print the variable *X* starting in column 5 and *Y* in column 21.

PRINT TAB (4); *X*; TAB (2Ø); *Y*

The variables are printed in the $n + 1$ position because the computer uses the first position for a sign. If negative it is displayed but when positive the position is left blank. It is important to note that the TAB function can only be used in the list following the PRINT keyword and that semicolons should be present. Examples of the use of the TAB function are found in Chapter 6 where simple graphics can be produced using this function.

Another formatting function is the SPC or space function that takes the form;

SPC (*n*)

where *n* is an integer number. This function allows *n* character positions to be skipped before text (or data) is printed. A title could be printed by the following statement;

```
1Ø PRINT SPC (16); "TITLE"
```

or

```
1Ø PRINT SPC (16); "TITLE"; SPC (8); "MY PROGRAM"
```

The SPC function can be used several times within a PRINT statement and counting of spaces to be skipped will be from the present print position when the SPC function is executed.

2.10 REM Statements

REM stands for 'remark' and it is a 'non-executable' statement which forms no part of the workings of a computer program. REM statements are useful as they are part of what is known as 'documentation'. They can be used to explain to a user what the program does and title the various sections of a program to make it understandable. Examples of REM statements can be seen in Program 2.7. It is good practice to start a program with several REM statements which give the program title, author and objectives.

2.11 END and STOP Statements

The END statement should be the last statement in a BASIC program since it marks the end of the program and is a signal to the BASIC interpreter and the computer's operating system. With microcomputers, END is often unnecessary but it is always good programming to include it.

The STOP statement has a slightly different function. A program can be written which, depending on certain conditions, does not need to go to the last statement and its execution could be stopped at that particular point. One way of doing this is to use the STOP statement. If desired, program execution can be continued after the STOP by use of the active command CONT.

2.12 Subscripted Variables (Lists and Arrays)

In engineering there are many instances

where tables and matrices are necessary. BASIC allows us to use 'subscripted' variables or arrays in these circumstances. Arrays are also invaluable when processing large amounts of data, particularly in the business field. Subscripted variables must be declared at the beginning of a program as the computer has to allow space within the memory before these variables can be stored or used. Subscripted variables can therefore be:

A (1Ø) — single dimension
B (1Ø,4) — two dimensions
A$ (12) — single dimension subcripted string variables

In fact subscripted variables can have multiple dimensions; they are not restricted to only two.

Declaration of subscripted variables is made using the DIM or dimension statement, for example:

1Ø DIM *A* (2Ø), *B* (1Ø,4), *A$* (3Ø)

In some versions of BASIC the number of elements in an array is greater than that declared, for example *A*(2Ø) could have 21 elements. The first element being *A* (Ø) and the last *A* (2Ø).

The handling of arrays in the computer will be examined in more detail in Chapter 4.

2.13 Program Control

A computer can perform calculations at great speed, to utilize this speed we need to be able to repeat similar calculations over and over again by *jumping* to different parts of a program. There are several methods of interrupting the flow of the program using specialized statements.

2.13.1 Unconditional Jumps

The normal sequence in which BASIC statements are obeyed can be changed by using a jump statement. In its simplest form it is an *unconditional* jump and the computer *must* obey it. The BASIC statement is the GOTO and takes the form

GOTO *XXX*

where *XXX* is a line number. It can be used to jump to any line in a program and execution of the program will continue from that line. If the GOTO points to a non-existent line number, the program will terminate with an error message being displayed. The simplest use is to enable repeat program runs. For example Program 2.5 when executed will only run once, however, repeated execution can be obtained by replacing line 16Ø by;

16Ø GOTO 4Ø

This will send the control back to line 4Ø after each calculation and the program will continue indefinitely. The only way such a program can be stopped is to press the 'Break' or 'Run/Stop' key.

2.13.2 Conditional Jumps

A conditional jump uses the IF statement and has the following format;

IF [condition] THEN [statement]

When the condition is met then the jump is executed and the statement is simply a GOTO. The condition is expressed using relational operators which are shown in Table 2.5

Table 2.5 Relational Operators

=	Equal
< >	Not equal
>	Greater than
> =	Greater than or equal to
<	Less than
< =	Less than or equal to

A typical example of an IF statement is as follows;

1ØØ IF *X* < = Ø THEN GOTO 5Ø

The condition can have either numbers, variables or expressions on both sides of the relational operator. Although the IF statement can be used to jump to different parts of a

Program 2.6

```
10 REM ******************************
20 REM     PROGRAM TO SHOW
30 REM     LOOPING USING COUNTER
40 REM     AND CONDITIONAL JUMP
50 REM ******************************
60 I=0
70 I=I+1
80 IF I>50 THEN STOP
90 Z=SQR(I)
100 PRINT I,Z
110 PRINT
120 GOTO 70
130 END
```

program by using a GOTO, it can also be used to alter variables or to output information or data, for example

10 IF $X < 0$ THEN PRINT "NUMBER TOO SMALL"

or

50 IF $(A+C) < (B+E)$ THEN $Z = A+B$

Both numeric and text variables can be used in the condition, but it is nonsense to compare a numeric variable with a text variable. When *text* is being compared, comparison takes place one character at a time and generally "Z" is considered to be greater than "A" and "9" is greater than "0". Thus "A" < "ABLE" < "DOG". A space is also less than any letter so that references or lists of objects can be sorted alphabetically.

2.13.3 Looping

Many programs will require certain statements to be repeated several times, such repetition is called 'looping'. Looping can be achieved by using unconditional jumps and setting up a counter. To demonstrate this let us consider a program to calculate the square root of the numbers from 1 to 50 (Program 2.6), which also demonstrates the use of the REM, STOP and END statements.

In Program 2.6, line 60 sets a counter *I* to zero which is increased by one in line 70. The following line (line 80) tests to see if *I* is equal to or greater than 50, if it is, then program execution is stopped. Line 90 then calculates the square root of *I* and the results are printed out by line 100. The unconditional jump from line 120 back to line 70 results in *I* being increased by one again. This jump back to line 70 will always occur until $I > 50$.

In this example we have set up our own counter and loop but there is a statement in BASIC that will do this for us. It is known as the FOR . . . NEXT statement. Program 2.7 is Program 2.6 rewritten using this statement and therefore shows it in its simplest form.

In Program 2.7, line 60, the FOR statement sets up a loop using *I* as a counter. The loop will be repeated with *I* being increased by one each time the loop is executed until it becomes greater than 50. The limit of the loop is indicated by the NEXT statement which, when encountered, causes the computer to increase the counter and then jump back to the start of the loop. In Program 2.7 the increment used is one by default, however, it is possible to change the increment by adding the STEP function to the FOR statement. For example, if line 60, Program 2.7, was altered to

60 FOR $I = 1$ TO 50 STEP 2

Program 2.7

```
10 REM ******************************
20 REM     PROGRAM 2:6
30 REM     RE-WRITTEN USING THE
40 REM     FOR....NEXT STATEMENT
50 REM ******************************
60 FOR I=1 TO 50
70 Z=SQR(I)
80 PRINT I,Z
90 PRINT
100 NEXT I
110 END
```

The program would then print the square roots of *odd* numbers only.

Many versions of BASIC allow the counter to start at any value and even permit decreases, for example the following is a valid FOR statement.

FOR *I* = 99 TO Ø STEP − 1

The start, end and increment values do not necessarily have to be numeric, they can also be expressions of any complexity.

FOR loops are important in engineering applications as many calculations involve one- or two-dimensional subscripted variables. In order to use these arrays FOR loops within FOR loops are a necessity. Loops within loops are called 'nested' loops and Program 2.8 illustrates how they are used to input and output subscripted variables.

Loop structures are easy to use and visualize but there are some rules and guidelines that should be followed, these are:

(a) Each nested loop must have a different index variable name (the counter variable).
(b) An inner loop, in nested loops, must be terminated *before* the outer loops.
(c) With every FOR statement there must be an associated NEXT statement to terminate the loop.
(d) In general a loop should be entered through its FOR statement and either:
    (i) Jump back at the NEXT statement; or
    (ii) Jump out, and *stay* out.
(e) Never jump into the middle of a loop.
(f) Try not to alter the counter within a loop or use the value of the counter outside the limits of the loop.

The most common mistakes with program loops are due to (b) and (c). A FOR . . . NEXT loop does not have to be completed. In certain calculations it may be necessary to terminate the loop when a specific condition is encountered: a conditional jump will allow the program to exit the loop and continue the program at another point.

2.13.4 Multiple Branching

Branching to a single point in a program is satisfactory for simple decisions, however, a more versatile branching statement is the computed GOTO which allows multiple branching. It takes the form;

ON variable or expression GOTO line number, line number, . . . for example,

ON A GOTO 2ØØ, 25Ø, 4ØØ, 7ØØ

In this example the value of variable *A* determines which line the computer jumps to. When *A* is 1 it jumps to line 2ØØ; when *A* is 2, line 25Ø, and so on. The variable *A* must be an integer but if it is real then the value used depends on the version of BASIC in the com-

Program 2.8

```
10 REM ******************************
20 REM DATA INPUT AND OUTPUT
30 REM FOR A TWO DIMENSIONAL
40 REM ARRAY
50 REM ******************************
60 DIM A(3,5)
70 FOR I=1 TO 3
80 PRINT "INPUT ROW ";I
90 FOR J=1 TO 5
100 INPUT A(I,J)
110 NEXT J
120 NEXT I
130 PRINT
140 PRINT" OUTPUT OF ARRAY"
150 FOR I=1 TO 3
160 FOR J=1 TO 5
170 PRINT A(I,J);
180 NEXT J
190 PRINT
200 NEXT I
210 END
```

puter. Some versions round the value to the nearest integer value, others take the integer part only.

2.14 Subroutines

A subroutine is a series of statements, almost a separate program, which can be repeated many times from within a main program. Subroutines save typing the same lines many times. In fact most programs can be written in the form of a series of subroutines which can be tested and corrected separately and then combined at a later stage.

Branching from the main program to a subroutine is accomplished by the GOSUB statement which takes the form:

GOSUB *XXX*

where *XXX* is the line number signifying the start of the subroutine. Once the program has branched to the subroutine, the end of the subroutine is signalled by the RETURN statement. The computer then branches back to the statement following the GOSUB. Program 2.9 illustrates the GOSUB statement. The subroutine is called ten times from a FOR loop. Subroutines can also be nested, that is a subroutine can call another subroutine which in turn could call a third, and so on.

Multiple branching to subroutines can be achieved by a *computed* GOSUB which has a similar format to the computed GOTO, for example:

ON A GOSUB 2ØØ, 25Ø, 4ØØ

This command may not be available in all versions of BASIC.

2.15 User Definable Functions

Subroutines usually contain several statements but in certain instances formulae can be calculated on one line. The user definable function is a more elegant way of programming this type of problem. Suppose we wish to calculate

Program 2.9

```
10 REM ******************************
20 REM     MAIN PROGRAM
30 REM ******************************
40 FOR I=1 TO 10
50 PRINT
60 GOSUB 100
70 NEXT I
80 PRINT "THIS IS THE END OF THE PROGRAM"
90 STOP
100 REM ** START OF SUBROUTINE **
110 PRINT "THIS IS CALL ";I
120 FOR J=1 TO 500
130 NEXT J
140 PRINT
150 RETURN
160 END
```

$x^2 + 3x - 4$

with varying values of x at different points in a program. It would be a waste of effort to type out the formula several times so we would define the function as follows:

DEF FNB (X) = X↑2 + 3*X − 4

The DEF FN defines the function B (any legal numeric variable can be used instead of B) and whenever this function is evaluated all that is needed is the statement FNB(X) where X is the value to be used in the equation. Program 2.10 demonstrates the use of this statement.

The variable name used within the DEF FN statement can be used elsewhere in the program without affecting the function. There are, however, other rules that must be obeyed, these are:

(a) The function defined must fit on one line of the program.

(b) When referencing the function a single

Program 2.10

```
10 REM ***************************
20 REM    DEFINABLE FUNCTIONS
30 REM ***************************
40 DEF FN B(X)=X↑2+3*X-4
50 FOR I=0 TO 3 STEP 0.25
60 Y=FNB(I)
70 PRINT I,Y
80 NEXT I
90 PRINT
100 FOR I=5 TO 30 STEP 5
110 Y=FNB(I)
120 PRINT I,Y
130 NEXT I
140 END
```

variable enclosed in parentheses must follow the function name.

(c) When defining the function, the variable name in parentheses is the only one that can appear to the right of the equals sign.

2.16 Commands

As stated at the beginning of the chapter some BASIC keywords are used to make BASIC carry out an operation when the command and control are passed back to the computer. We have already mentioned the RUN command which executes a program but there are others.

2.16.1 Program Listing — LIST

The LIST command is quite specific, it causes the computer to list all the program lines in sequence. Listing a complete program is useful for checking its correctness, however, the LIST command can have numeric parameters. For example:

LIST 5Ø

will print out only line 5Ø, but in some versions of BASIC it would print out line 5Ø and all subsequent lines. Typing

LIST 5Ø - 2ØØ

or

LIST 5Ø, 2ØØ

would cause all the lines from 5Ø to 2ØØ to be printed.

2.16.2 Clearing Computer Memory — NEW

Once a program has been written and used and is no longer required then it should be erased from the memory. This is accomplished by the NEW command. Once executed the program which was in memory can no longer be accessed. Failure to type NEW can cause a new program that is entered to be mixed with the old one. If a program is required at a later date it should be stored in backing store by the SAVE command prior to entering NEW.

2.16.3 Program Storage — SAVE

The SAVE command instructs the computer to transfer the program in memory to a program file (see Chapter 3) in its backing store of either magnetic tape or disks. The SAVE command has the general format:

SAVE [program name]

Some versions of BASIC require the program name to be in quotation marks, others (especially mainframe computers) use the name of the work area into which a program is written. The program name becomes the name of the program file on backing store. All versions require every program file to have a different name and an attempt to store a program in a file that already exists will cause an error message to be displayed.

2.16.4 Program Retrieval — LOAD

Retrieval of programs from backing store can be achieved by the LOAD command and has a similar format to SAVE, for example:

LOAD [program name]

Execution of the LOAD command will cause the computer to clear its memory prior to loading the requested program, but it is wise to clear the memory first using the NEW command. Some versions of BASIC use OLD instead of LOAD and in others there are special versions such as DLOAD which means load the program from disk backing store.

There are many additional commands that can renumber programs and join programs together but these are not found on all computers and they vary in detail from machine to machine.

2.17 Typing and Editing Programs

Most versions of BASIC allow line lengths of

up to 80 characters, but whatever number is allowed a warning will be signalled if the number is exceeded.

When typing BASIC statements it is good programming practice to use the space bar to separate out keywords. It may be necessary in some versions of BASIC and it does make programs more readable.

The majority of programmers never write a program which works perfectly the first time and logic and typing errors have to be corrected. This is known as 'editing' or 'debugging' a program. Editing facilities on computers vary tremendously but there is one method that is common to all versions of BASIC. It is a procedure known as 'overwriting'. An incorrect line can be corrected by simply retyping the whole line again. The computer will accept it and replace the old line with the newer version.

An unwanted line can be erased from the program by entering just the line number and leaving the rest of the line blank. When the computer receives a blank line, it erases it from its memory.

New lines can be added at any time by typing the new line but one must make certain that the line number for the new line does not already exist in the program. For this reason it is wise to number the lines in gaps of 1Ø or 1ØØ to enable additional lines to be entered as the program develops. The first line number should be greater than zero but the upper limit depends on the computer and the manuals should be consulted.

Further Reading

Alcock, D. *Illustrating Basic*. Cambridge University Press, 1977.

Basic Programming Reference Manual (Applesoft). Apple Computer Inc., 1981.

Bishop, P. *Further Computer Programming in Basic*. Nelson, 1982.

Gosling, P. E. *Beginning Basic*. MacMillan, 1977.

Gosling, P. E. *Continuing Basic*, MacMillan, 1980.

Lein, D. A. *The Basic Handbook*. Compusoft Publishing, USA, 1978.

Osborne, A. and Donahue, C. *Pet/CBM Personal Computer Guide*. Osborne/McGraw-Hill, 1980.

3

More Advanced BASIC and the Microcomputer

3.1 Introduction

The 'low cost' computer has had a dramatic effect on the development of BASIC. The language has been expanded to include many extra commands enabling its use in every sphere of business. In particular it has been enhanced to deal with text and text variables allowing word processing and sorting to be accomplished with ease. This chapter will deal with some of the commands found in microcomputer BASIC, together with a brief survey of a microcomputer's storage and file handling abilities. Due to the large number of different microcomputers available there will only be a general coverage of storage and files.

3.2 Microcomputers

A microcomputer is almost impossible to define precisely due to the large number of types available. Most people visualize them as scaled down versions of larger computers using a typewriter-style keyboard for input and a VDU or ordinary television screen for output. The computer's operating system, BASIC interpreter and memory are in small integrated circuits on one or more electronic boards inside the keyboard case. Their simplicity of operation enables a user to produce programs in a relatively short time. To understand their full capabilities requires a little knowledge of how they store information (programs and data). For example, a potential buyer will often be told that a particular machine is capable of storing 32 kilobytes, but what are 'bytes'? To explain this we need to examine binary numbers and computer storage. Although a knowledge of binary numbers is not essential to the understanding of computer storage it is particularly useful when microcomputers are interfaced with test equipment.

Whereas the decimal number system is to the base 10, the *binary* system has only two digits, zero and one, and is therefore to the base 2. It is used in computers because it is easy to tell if electrical devices and switches are ON (one) or OFF (zero). Electrical and magnetic storage devices can also be constructed having two stable states, thus the binary system is ideal. Binary numbers are constructed just like decimal numbers. For example, decimal 562 can be constructed as follows:

$$562 = (5 \times 10^2) + (6 \times 10^1) + (2 \times 10^0)$$

Binary numbers can also be formed in a similar manner, for example binary 101 is constructed as follows:

$$101 = (1 \times 2^2) + (0 \times 2^1) + (1 \times 2^0)$$

Thus binary 101 is the same as decimal 5. A single binary digit (0 or 1) is known as a *bit* and all computers use groups of these. The majority of microcomputers use a group of 8 bits which is known as a *byte*, but there are now some using 16 or 32 bits.

A byte of 8 bits has the following limits: 00000000 and 11111111. In decimal numbers this is 0 to 255. A microcomputer, therefore, works with bytes having values between 0 and

255. All programs and data are stored in the form of binary numbers but the CPU, the operating system and the BASIC interpreter can identify and distinguish between data and programs.

The storage capabilities of microcomputers are expressed in terms of the number of bytes. One kilobyte of storage is in fact 2^{10} bytes (1024 bytes). A machine having a storage capacity of 32K, therefore, has 32 768 storage locations, each location capable of storing 1 byte. An 8 bit CPU such as the 6502 found in Apple and Commodore microcomputers can address up to 64K bytes of memory. Not all of this memory is used to store programs.

3.2.1 ROM — Read Only Memory

The part of the computer's memory which is permanently programmed by the manufacturer is known as ROM. It cannot be changed by a user; it can only be read *from*, hence, *Read Only* Memory. ROM contains all the program code for the computer to function, the operating system and, if the machine runs in BASIC, the BASIC interpreter. Some microcomputers have empty sockets on their printed circuit boards into which additional ROMs may be plugged allowing the capabilities of the computer to be increased at a later date.

3.2.2 RAM — Random Access Memory

This is the area of memory in which programs and data are stored and, therefore, can be written *to* and read *from*. When power to a microcomputer is switched off anything stored in RAM is normally lost and is often referred to as 'volatile' memory. A section of RAM in all microcomputers is reserved by the operating system to store information it requires. An area of memory is also set aside and used for outputting to the VDU. Anything output to the VDU is stored in this area; video scanning routines scan the memory area displaying the contents on the screen. The screen display is said to be *memory mapped*.

3.2.3 Interfaces

Microcomputers are provided with a variety of sockets or interfaces which enable the computer to exchange information with other electronic devices (peripherals), for example cassette players, printers, disk units or laboratory test equipment. All input and output to these devices are again memory mapped and require a small area of reserved RAM.

This description of a microcomputer is very brief and simplified but certain aspects will be covered in greater detail in Chapters 6 and 12.

3.3 Additional BASIC Commands Found on Microcomputers

The additional BASIC commands that have been developed relate to one of three areas:

(a) string variables: improved handling facilities;
(b) storage and retrieval of programs;
(c) graphics.

In both (b) and (c) the commands are machine dependent and cannot be transferred from computer to computer. Commands associated with (a) are almost standard but there may be subtle variations from one computer to another.

As defined in Section 2.4, a string is a line of text and can consist of alphabetic, numeric or graphics characters in any combination. In certain circumstances a string may be entered or defined during a program run and remains unchanged until the program terminates. In other instances the string may be repeatedly changed. To allow a user to manipulate strings, several string functions are now available on the majority of microcomputers. Many of the functions will not be used by beginners, new to BASIC, yet they allow powerful data vetting routines to be written for advanced programs.

Program 3.1

```
10 REM ******************************
20 REM    GET STATEMENT
30 REM  COMMODORE PETS,& APPLE II
35 REM REMOVE LINE 90 FOR APPLES
40 REM ******************************
50 PRINT
60 PRINT"ARE YOU OLDER THAN 25 YEARS"
70 PRINT"TYPE Y OR N"
80 GET A$
90 IF A$="" THEN GOTO 80
100 IF A$="Y" THEN GOTO 130
110 IF A$="N" THEN GOTO 150
120 GOTO 70
130 PRINT" YES  : AGE > 25"
140 STOP
150 PRINT" NO   : AGE < 25"
160 END
```

3.3.1 Single Character Input

Many versions of BASIC have a statement which allows single characters to be input from the keyboard, but unlike the INPUT statement they do not halt the execution of the program waiting for the ENTER or RETURN key to be pressed.

In Commodore BASIC and Applesoft BASIC the command takes the form

GET [variable]

and can be used to accept either a single numeric value or a single character, for example

GET *A* : accepts a digit only
GET *A$* : accepts a single character (alphabetic, numeric or graphic)

Although the format is the same in both versions of BASIC there is a subtle difference in operation. In Applesoft BASIC, on execution, the GET statement causes the microcomputer to wait until the key is pressed, with Commodore BASIC it only checks to see if a key has been pressed. If no key has been pressed it automatically moves on to the next statement, thus in order to wait for a key depression a loop must be set up.

In TRS 80 and Video Genie BASIC the statement takes the form

INKEY$

and accepts characters only.

These statements are particularly useful for testing the depression of certain keys, especially after a question has been asked from a program. Programs 3.1 and 3.2 demonstrate how both versions of these statements can be used. The programs simply ask the users if they are older than 25 years of age and to respond by typing the key Y or N. The statement in lines 9Ø and 1ØØ tests for a null character and is checking if a key has been pressed or not. If no key has been depressed there is a null return and the program loops back to check again. This is repeated until a key is pressed. Once a key is pressed lines 1ØØ and 11Ø check to see if it was either Y or N and to take the appropriate action. The depression of any other key will cause the program to loop back to line 7Ø repeating part of the instructions again.

These statements can also be used to *concatenate* strings. The arithmetic 'plus' operator *adds* the contents of numeric variables but

Program 3.2

```
10 REM ********************************
20 REM        INKEY$ STATEMENT
30 REM  FOR TRS-80 & VIDEO GENIE
40 REM ********************************
50 PRINT
60 PRINT"ARE YOU OLDER THAN 25 YEARS"
70 PRINT"TYPE Y OR N"
80 PRINT
90 A$=INKEY$
100 IF A$="" THEN GOTO 90
110 IF A$="Y" THEN GOTO 140
120 IF A$="N" THEN GOTO 160
130 GOTO 70
140 PRINT" YES  : AGE > 25"
150 STOP
160 PRINT" NO   : AGE < 25"
170 END
```

when used with strings it simply *joins* (concatenates) them together. Suppose we let

A$ = "FRED "

and

B$ = "SMITH"

the following statement would concatenate them, assigning "FRED SMITH" to *C$*.

C$* = *A$ + *B$*

Only the *plus* operator will work, the – , *, / operators are invalid *string* operators. The use of single character input combined with concatenation can be used by programs to allow passwords to be entered which are not displayed (Program 3.3). Another use is for halting output to a VDU so that any information already displayed can be read by the user at leisure. Continuation of the program can then be achieved by a single key depression.

3.3.2 String Handling Functions

String handling functions which are available can again have subtle variations from machine to machine. A general description of the functions will be given but examples will be specific to Commodore BASIC and Applesoft BASIC. The functions available are:

LEN — Returns the number of characters contained in a string of text or text variables placed within parentheses following the function, for example,

LEN ("*ABCDEF*")

or

LEN (*A$*)

where text has been previously assigned to *A$*.

LEFT$ — Returns the left part of a text string. The function's argument gives the number of characters returned, for example,

LEFT$ (*B$*, 5)

returns the five leftmost characters of the string in text variable *B$*.

RIGHT$ — Returns the right part of a text string; the number of characters returned depending on the function's argument, for example,

Program 3.3

```
10 REM ********************************
20 REM  CONCATENATION OF STRINGS
30 REM  COMMODORE PET'S
40 REM ********************************
50 A$=""
60 P$=""
70 REM *****ENTER A 5 CHARACTER WORD
80 FOR I=1 TO 5
90 GET A$
100 IF A$="" THEN 90
110 P$=P$+A$
120 NEXT I
130 PRINT
140 PRINT"PASSWORD IS ";P$
150 END
```

RIGHT$ (*B$*, 5)

returns the five rightmost characters of the string in text variable *B$*.

MID$ — Extracts a section of a text string by specifying the leftmost starting position and the number of characters to be returned. For example,

MID$ (*A$*, 3, 6)

returns six characters starting from the third in the text string associated with *A$*. One must ensure that the starting position is within the text string. If the number of characters to be returned is absent then all characters to the right of the starting position are returned.

STR$ — Converts a number to its equivalent string of text characters.

VAL — Converts a string of text characters to its equivalent number. If the first character is not numeric then a zero is often returned.

CHR$ — Converts a numeric ASCII code to its equivalent text character. ASCII stands for American Standard Code for Information Interchange. All keys have an ASCII code, for example, *A* is 65 and *B* is 66. Some keys do not print anything on the screen but they still have an ASCII code. The return key is an example having the ASCII code of 13 (see Chapter 6).

ASC — Is the reverse of CHR$, it returns the ASCII code (in decimal) for the first character of the specified string.

These functions allow text to be manipulated and validation routines to be written and incorporated in a program. Validation routines are useful in attempting to make programs 'foolproof'. A routine which ensures that only a numeric number is entered is shown in Program 3.4. The program accepts characters one at a time and checks them. If it is not a numeric character it is not concatenated to the input string. The computer will ignore all characters other than 0 – 9, a full stop (decimal point) and a minus sign; and it will wait for the next character. These checks are carried out in lines 11Ø to 14Ø.

Program 3.4

```
10 REM ***********************************
20 REM    NUMERIC INPUT ROUTINE
30 REM COMMODORE BASIC
40 REM APPLESOFT BASIC
50 REM ***********************************
60 PRINT
70 PRINT"ENTER A NUMBER"
80 GETA$
90 IF A$="" THEN GOTO 80
100 IF A$=CHR$(13) GOTO 190
110 IF A$="." THEN GOTO 150
120 IF A$="-" AND P$="" THEN GOTO 150
130 IF A$<"0" THEN GOTO 80
140 IF A$>"9" THEN GOTO 80
150 P$=P$+A$
160 IF LEN(P$)=10 THEN GOTO 190
170 PRINTA$;
180 GOTO 80
190 PRINT
200 PRINT
210 P=VAL(P$)
220 PRINT"NUMERIC VALUE P IS  ";P
230 END
```

Only ten numeric characters are allowed in the routine and line 160 checks this using the LEN function. If ten characters have been entered the routine is terminated by jumping to line 190 where the character string is converted to a number which is then printed out at line 220. Although a maximum of ten characters is allowed, fewer than ten can be entered. This is achieved by testing for the depression of the return key using the CHR$ function in line 100. The routine in Program 3.4 does have its limitations but it is left for the reader to improve it.

It should also be reaffirmed that strings can be dimensioned, for example,

DIM *A$* (12)

In Commodore BASIC and Applesoft BASIC this statement would allow 12 strings to be stored one in each element of the array, with each element capable of handling a string of up to 256 characters.

Program 3.5 demonstrates the other string-handling functions to check if a valid month is entered. The lines 60 to 100 enter the required data into the program. The months are read as two text strings into the variables *M1$* and *M2$* which are concatenated to form one string *M1$* in line 200. Lines 120 to 180 accept three characters from the keyboard. The three characters are then compared with the month abbreviations in string variable *M1$* using the MID$ function in the FOR loop; lines 210 – 260. If there is a match then statement 240 prints out the appropriate right-hand portion of the data in variable *O$*. If no match is found then *O$* is printed out in full. This method of checking text is one that can be used to select different parts of a program from several options.

3.4 PEEK and POKE

Every storage location in a computer's

Program 3.5

```
10 REM ********************************
20 REM  STRING HANDLING FUNCTIONS
30 REM COMMODORE BASIC
40 REM VALIDATION OF CORRECT MONTH
50 REM ********************************
60 DATA JANFEBMARAPRMAYJUN
70 DATA JULAUGSEPOCTNOVDEC
80 DATA INVALID MONTH ENTERED
90 READ M1$,M2$
100 READ O$
110 PRINT
120 PRINT"ENTER MONTH : ONLY FIRST 3 CHARACTERS"
130 FOR I=1 TO 3
140 GET A$
150 IF A$="" THEN 140
160 M$=M$+A$
170 PRINTA$;
180 NEXT I
190 REM *** CONCATENATION OF MONTH STRINGS
200 M1$=M1$+M2$
210 FOR I=1 TO 36 STEP 3
220 IF M$<>MID$(M1$,I,3) THEN GOTO 260
230 PRINT
240 PRINT MID$(O$,3,12);M$;RIGHT$(O$,8)
250 GOTO 310
260 NEXT I
270 PRINT"********************************"
280 PRINT O$;"   ";M$
290 PRINT"********************************"
295 M$=""
300 GOTO 120
310 PRINT
320 PRINT" END OF ROUTINE"
330 END
```

memory has an address (a number) which is unique, and in an 8 bit microcomputer each location can store a number in the range 0 to 255. Programs and data are stored as series of numbers in the memory locations. Most versions of BASIC provide two functions allowing a single memory location to be examined and in some instances (with RAM memory) changed. The functions are normally called PEEK and POKE.

The PEEK function examines any memory location and takes the format

PEEK (*A*)

where *A* is a memory location. The value returned is the decimal value of the content of memory location *A*. The value will correspond with a character or BASIC command which is obtained from lists provided in the microcomputer's manuals.

The POKE function enables a value (between 0 and 255 for 8 bit microcomputers) to be placed in a specified memory location and takes the format;

POKE A, B

where A is the memory location and B is the value to be placed in memory location A.

The PEEK can be used on both RAM and ROM, but POKE can only be used on RAM memory. These functions should be used with care, indiscriminate poking can cause the computer to crash and lose everything from RAM memory. PEEK and POKE functions can be used effectively to provide graphic displays; see Chapter 6.

3.5 Miscellaneous Additions

Among the additions to BASIC which can be classed as miscellaneous, there are two worthy of mention. The first is the ability to print a message using an INPUT statement, for example,

```
INPUT "ENTER X VALUE"; X
```

This single statement suffices for both a PRINT and INPUT statement; the message in quotes being printed just as it would be with a PRINT statement.

The second addition worthy of mention is the acceptance of more than one statement per line number, for example,

```
100 IF X > Y THEN Z = X - Y :
GOTO 300
```

Here the colon (:) is used to separate the two statements. This feature makes for economical use of memory, however, one must be careful in its use. In the example given if $X < Y$ then the GOTO statement would *not* be executed. This may be the intention of the programmer but its use should be planned.

Both of these features make programs more difficult to read and, therefore, in the examples in this text an attempt has been made to use them sparingly.

3.6 Backing Store

Programs and data can be stored on a magnetic medium which is external to the microcomputer and takes the form of either an *audio cassette* or a *diskette*. A diskette (or disk) stores information in the same way as a cassette, the biggest difference being that a disk is a thin round piece of vinyl coated in a magnetic material. The disk spins like a record and the *read* and *write* (record/playback) head is moved across the disk to read or write information as the disk spins. The diskette is sealed inside a paper cover with a special coating on the inside to remove dust from the diskette itself. The read/write head is allowed close to the diskette at a predefined window cut in the paper sleeve. The advantage of a diskette is that it can access large amounts of data faster than a cassette but there is of course a price difference.

Backing store devices must be controlled by the computer and, therefore, there are special operating routines enabling data and programs to be transferred and stored on these devices. The routines are known as Cassette Operating Systems (COS) and Disk Operating Systems (DOS). Both systems have commands that can be entered from the keyboard or written into a BASIC program and thus the commands can be considered to be extensions of the BASIC language. A problem arises because there are no standard definitions and formats for these commands. Also programs stored on a disk, say, by an Apple II computer, cannot be read by a Commodore computer without special equipment and, therefore, portability of data and programs is only between like machines. Cassette operating systems are normally standard and built into even the simplest machines. A disk operating system can, however, be either in the form of firmware or software. For example, Commodore computers have disk-operating commands and routines in ROM; Apple II computers require the system to be loaded in from disk before any disk commands can be issued. The DOS in this case is stored in a protected area of

RAM and is, therefore, an extension to the system which can be added at any time.

We consider now how and where data are stored on cassettes and disks. On cassettes pieces of information are recorded one after another in a sequence and there is a beginning and an end to the sequence of recorded information. In order to obtain data from the middle of such a sequence it must be read starting from the beginning, it is impossible to start it part way through. With diskettes the information is stored on a number of concentric tracks. The tracks vary in number from machine to machine. An Apple II (with DOS 3.3) and a Commodore 4040 disk unit are formatted in 35 tracks but the Commodore 8050 disk unit has 77 tracks. Each track is divided up into sectors, Fig. 3.1, and one sector will cover all tracks from the edge of the diskette to its centre. Data are only recorded on a track within a sector. The reason being that if data were recorded over the whole length of a track, then every track would hold different amounts of data as the length of a track reduces the closer it is to the centre of the diskette. Each sector will hold a specific number of bytes, on 8 bit microcomputers it is 256 bytes. The majority of disk drives write the same number of sectors on the outer track as on the inner one, but Commodore disk drives write more sectors on the longer tracks.

The disk-operating system normally uses two tracks to keep a record of the information stored and where it is. The record is in the form of an index and referred to as the 'Directory' or 'Catalog', which can be examined to find out what has been stored on the disk. Because the directory records where everything is, the stored data can be accessed quickly without having to search through everything that has been stored. It is this ability which makes them much more useful for storing data than a cassette. All disks prior to being used must be *initialized*. The disk is formatted and the directory tracks are set up during initialization.

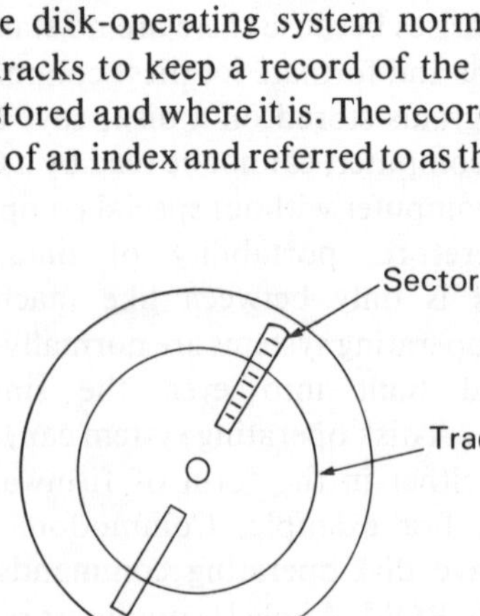

Fig. 3.1 Diskette.

3.7 The File Concept

A logical system is needed to distinguish between stored programs and data used by programs. This is done by storing information in 'files'. By definition a file is an organized collection of data. To obtain information from a file it must be 'opened' and will remain so while it is being used, after which it is 'closed'. Files can be of two types: program files or data files.

3.7.1 Program Files

A 'program' file contains a listing of a program, i.e. the BASIC program statements. A program file is created every time the SAVE command is used. It can be read and the program loaded back into the computer by the LOAD command. The maximum size of a program file is limited to the memory size of the computer that created it. A 16K computer can only run programs that are less than 16K in length.

3.7.2 Data Files

Data files contain data that can be read and used by a program. The size of a data file is only limited by the capacity of the storage medium. The information put into a data file can be divided into units known as records and each record can be further subdivided into fields (Fig. 3.2). Beginners to programming usually have only one field per record for simplicity but as they become more proficient so the structure of their records

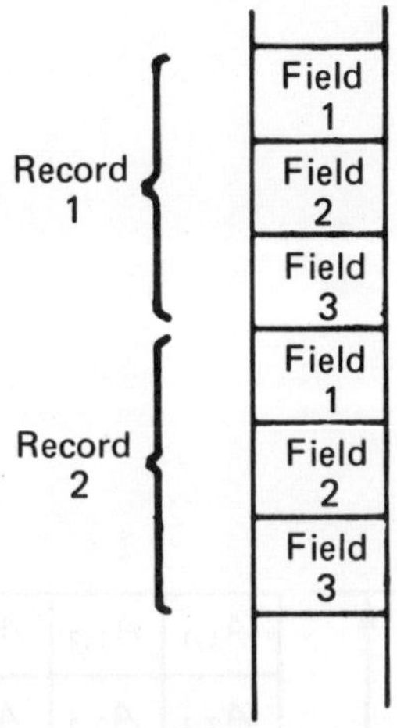

Fig. 3.2

develops. Records with several fields are generally used when data are repetitive.

Fields and records can be illustrated by considering the data in Table 3.1. The table contains information which could be used to describe four soil samples. All the data for one soil sample can be referred to as a 'record' and the individual values, 'fields'. Thus, there would be six fields per record, Table 3.2. Notice that the data can be numeric or string depending on how the field has been defined.

Data, in the form of records, can be stored on a diskette in two ways and on a cassette tape in only one way. A cassette tape stores records one after another in a sequence, the file is called a 'sequential file'. The sequential file, or serial file, can also be formed on a disk. To access the twelfth record on a sequential file one must start at the beginning of the file and read records 1 to 11. The twenty-first record of the same file could then be obtained by reading forward from the twelfth, but what happens if we now, after reading the twenty-first record, require the eighteenth. The only way this can be done is to go back to the start of the file and read the records 1 to 17 until we then arrive at the eighteenth. In some versions of BASIC it may even be necessary to CLOSE the file first and then re-OPEN it.

To overcome this direct access data files can also be created. Direct access files (or 'random' files) allow a greater flexibility than sequential files but they are more difficult to program. Versions of BASIC which support direct access files provide special statements allowing individual records to be accessed.

The method of OPENing files, naming them, storing the data, separating the records and fields and using them tends to vary from computer to computer. The reader is urged to examine the manuals of the computer he is using to familiarize himself with the procedures.

Table 3.1

Sample no.	Liquid limit	Plastic limit	Natural moisture content	Unit wt.	Soil type
1	73	25	30	17.3	CH
2	68	23	53	14.8	ML
3	46	24	18	18.0	ML
4	25	15	17	19.0	CL

Table 3.2

		Field
R	1	Sample number
E	2	Liquid limit
C	3	Plastic limit
O	4	Moisture content
R	5	Unit weight
D	6	Soil type

Further Reading

Osborne, A. and Donahue, C. *Pet/CBM Personal Computer Guide*. Osborne/McGraw-Hill,

Poole, L. McNiff, M. and Cook, S. *Apple II User's Guide*. Osborne/McGraw-Hill, 1981.

The DOS Manual (Disk Operating System). Apple Computer Inc., 1981.

4

Computing with Arrays

4.1 Introduction

Many instances arise in civil engineering when it is convenient to store information in ordered sets called 'arrays'. Usually the stored information is in numeric form but alphabetic arrays are also acceptable in BASIC. Arrays are usually one-dimensional or two-dimensional but higher orders are possible subject only to storage limitations in the computer.

As an example of a one-dimensional array consider the representation of the four items of data which might be needed to perform a structural analysis of a beam: I = node number at end 1; J = node number at end 2; L = length; EI = flexural rigidity.

This information could be stored in a one-dimensional array which we might call 'A'. Thus,

$A(1) = I$
$A(2) = J$
$A(3) = L$
$A(4) = EI$

which we can imagine to be stored in the computer as

$A(1)$	$A(2)$	$A(3)$	$A(4)$
I	J	L	EI

If we wish to use this method for storing the properties of a number of beams, then we can adopt a two-dimensional array in which each *row* of four elements relates to *one* of the beams. If there are N beams to represent, the

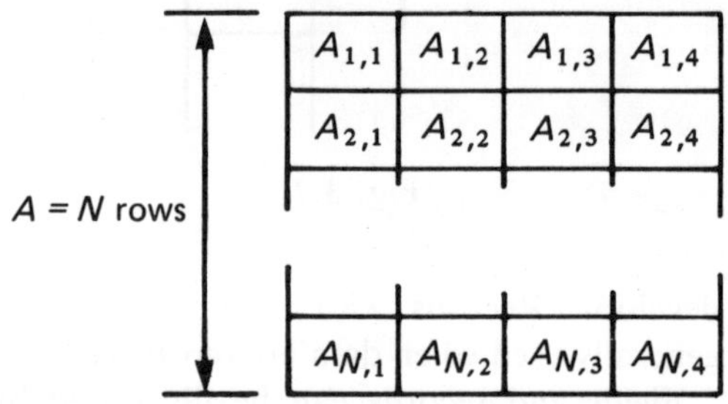

Fig. 4.1

array will have N rows and four columns, as in Fig. 4.1.

If we retain the name 'A' for this array then A is an ($n \times 4$) array and if we wish to 'address' the elements of the array we use a double subscript notation as $A(i, j)$ where i denotes the row and j the column of A. For example if $i = 7$ and $j = 3$ we are addressing the location which contains the length of beam number 7. The similarities with matrix algebra are very clear. Indeed 'array' is just another word for 'matrix'.

In this chapter we shall be concerned with the processing of arrays. This includes the operations of dimensioning, input, output, addressing and arithmetical operations. We shall also look, in Section 4.6, at the process of 'sorting' the contents of arrays.

4.2 The Dimension Statement

The general form of the dimension statement in BASIC is:

line number DIM name (dimension 1, dimension 2, . . .)

Examples are:

(a) 200 DIM *B* (*NB*, 11) defines '*B*', $NB \times 11$ elements

(b) 270 DIM *K* (*N*, *N*), *R* (*N*), *D*(*N*) defines
'*K*', $N \times N$ elements
'*R*', *N* elements
'*D*', *N* elements

(c) 320 DIM *C* ($N+Q$, 9) defines '*C*', $(N+Q) \times 9$

The above examples contain both one- and two-dimensional arrays. The number of dimensions is limited only by the storage available. Frequently the dimensions of the arrays are variables such as *NB* and *N* in (a) and (b) and ($N + Q$) in (c). This allows a program to be written in general terms so that the dimensions of the arrays can be input with the data relating to a particular problem to be solved. Some versions of BASIC do not include this facility. In the DIM statement it is assumed that the array dimensions start from zero. This effectively increases the dimensions in the above examples by one; the corresponding storage locations need not be used but are available if required.

Arrays can be dimensioned at any stage in a program but cannot be redimensioned unless a CLEAR statement is introduced. It is not usually necessary to redimension arrays during the execution of a program. If variables are used in array dimensions, these variables must have been assigned numerical values *before* the DIM statement is encountered. Usually the variable dimensions, such as *NB*, *N* and *Q* in the above examples, are items of data to be input during execution of the program.

4.3 Input and Output of Arrays

Arrays may be generated within the computer during program execution, for example.

```
100 DIM A (N,M)
110 FOR I = 1 TO N
120 FOR J = 1 TO M
130 A (I,J) = arithmetical expression
140 NEXT J
150 NEXT I
```

will generate an $N \times M$ array; the control index *I* denoting the row number being generated and *J* denoting the column number. The array is generated row-by-row as will be evident if the two loops are studied. The arithmetical expression will normally involve functions of *I* and *J*.

The usual method of inputting an array is by the use of the INPUT statement. This is in the form,

line number INPUT item list.

The array *A* ($N \times M$) could be input by,

```
180 FOR I = 1 TO N
190 FOR J = 1 TO M
200 INPUT A (I, J)
210 NEXT J
220 NEXT I
```

This will read in the array *A* with the elements, each separated by RETURN, arranged row by row. For example, if

$$A = \begin{bmatrix} 1 & 2 & 3 \\ 4 & 5 & 6 \end{bmatrix}$$

The data will be typed in,

1 RETURN
2 RETURN
3 RETURN
etc.

This would not be a very satisfactory program for other than a very small array. With larger arrays, it would be preferable to build in more control over the input process. Program 4.1, using strings, would be an improvement for the input of an array a ($N \times P$). This program will input the matrix row-by-row with SPACE between the elements in each row.

Arrays may be displayed using the PRINT statement. To display matrix *A* ($N \times M$) we could use Program 4.2 which displays the matrix column by column with a pause

Program 4.1

```
10 REM THIS PROGRAM WILL INPUT AN ARRAY A (NXP) ROW-BY-ROW
20 REM INSERT 'SPACE' BETWEEN NUMBERS
30 PRINT "TYPE IN NUMBER OF ROWS"
40 INPUT N
50 PRINT"TYPE IN NUMBER OF COLUMNS"
60 INPUT P
70 DIM A(N,P)
80 FOR I = 1 TO N
90 PRINT
100 PRINT"ENTER ROW ";I;" 'SPACE' BETWEEN NUMBERS"
110 FOR J = 1 TO P
120 GET A$
130 IFA$=""THEN GOTO 120
140 B$=B$+A$
150 PRINT A$;
160 IF A$=" "THEN GOTO 180
170 GOTO 120
180 A(I,J)=VAL(B$)
190 IF B$=" " THEN PRINT"0 ";
200 B$=""
210 NEXT J
220 PRINT
230 PRINT
240 NEXT I
250 END
```

Program 4.2

```
240 REM TO OUTPUT A RECTANGULAR MATRIX A(NXP)
250 FOR J = 1 TO P
260 PRINT "COLUMN";J;"OF MATRIX A"
270 FOR I = 1 TO N
280 PRINT A(I,J)
290 NEXT I
300 PRINT "PRESS ANY KEY TO CONTINUE"
310 GETA$: IF A$="" THEN 310
320 NEXT J
```

Program 4.3

```
10 DIM A(5),B(5)
20 FOR I = 1 TO 5
30 A(I)=I:B(I)=I↑2
35 NEXT
40 PRINT TAB(0)A(1)TAB(6)A(2)TAB(12)A(3)TAB(18)A(4)TAB(24)A(5)
50 PRINT TAB(0)B(1)TAB(6)B(2)TAB(12)B(3)TAB(18)B(4)TAB(24)B(5)
60 END
```

between each column controlled from the keyboard. Statement 31Ø is written in PET BASIC; in TRS 80 BASIC one would write,

```
31Ø A$ = INKEY$: IF A$ = " " THEN
31Ø
```

The display format can be controlled using the PRINT TAB statement. For example Program 4.3 will display,

```
1  2  3   4   5
1  4  9  16  25
```

To output an array to a printer, PRINT is replaced by LPRINT.

4.4 Addressing Arrays

In programming operations on arrays, usually the most difficult part is correctly to compute the 'addresses' of the array elements involved in each stage of the operation. The process is sometimes called 'indexing'. The problems involved are best illustrated with an example. Suppose we have a (4 × 4) matrix K in the computer and we wish to reduce this to upper triangular form using straightforward Gaussian elimination. The following program segment will achieve this:

```
2Ø FOR I = 1 TO 3
3Ø FOR J = I + 1 TO 4
4Ø D = K(J, I)/K(I, I)
5Ø FOR H = I TO 4
6Ø K(J, H) = K(J, H) - D * K
(I, H)
7Ø NEXT H
8Ø NEXT J
9Ø NEXT I
```

It will be observed that three variables (I, J and H) are used to 'index' or 'address' the elements of K used in the computation. In writing programs of this kind, it is useful to check the indexing on an example. Suppose we have the following matrix in the computer

$$K = \begin{bmatrix} 2 & -1 & 0 & 0 \\ -1 & 2 & -1 & 0 \\ 0 & -1 & 2 & -1 \\ 0 & 0 & -1 & 2 \end{bmatrix}$$

The reduction process may be traced in tabular form, Table 4.1. This program might now be generalized by introducing the variable N as defining the size of the matrix to be reduced. Statement number 2Ø would now become

```
2Ø FOR I = 1 TO N - 1
```

and the remaining modifications are left as an exercise for the reader.

It would not normally be necessary to complete the whole of a table like Table 4.1. Once the program has been written it can be tested on a known situation and the computer itself used to help the program development.

4.5 Matrix Operations

Most versions of BASIC provide for the standard operations of addition, subtraction, transposition, multiplication and inversion of matrices along with a facility for solving sets of linear equations. The two matrix programs available for PET provide the following matrix facilities:

MATRIX 1 (maximum dimension = 10)

E	enter	*	multiply
+	add	T	transpose
–	subtract	Q	quit
D	determinant	M	move
Y	display	?	list codes
R	read tape	C	change
		W	write tape

MATRIX 2 (maximum dimension = 15)

E enter
I inverse
Y display
C change
S simultaneous (coefficient matrix augmented by RHS)
Q quit
? list codes

Standard matrix operations on microcomputers are often quite slow and can be

Table 4.1 Trace of Indexing in Triangular Reduction of (4 × 4) Matrix

I	*J*	*D*	*H*	*K* (*J*,*H*)	*K* (REDUCED)			
1	2	$-\frac{1}{2}$	1	$-1-(-\frac{1}{2})(2)=0$				
			2	$2-(-\frac{1}{2})(-1)=1.5$				
			3	$-1-(\frac{1}{2})(0)=-1$				
			4	$0-(-\frac{1}{2})(0)=0$	2	−1	0	0
	3	0	1	0	0	1.5	−1	0
			2	−1	0	−1	2	−1
			3	2	0	0	−1	2
			4	−1				
	4	0	1	0				
			2	0				
			3	−1				
			4	2				
2	3	$-\frac{2}{3}$	2	$-1-(-\frac{2}{3})(1.5)=0$				
			3	$2-(-\frac{2}{3})(-1)=1.333$				
			4	$-1-(-\frac{2}{3})(0)=-1$	2	−1	0	0
	4	0	2	0	0	1.5	−1	0
			3	−1	0	0	1.333	−1
			4	2	0	0	−1	2
3	4	$-\frac{3}{4}$	3	$-1-(-\frac{3}{4})(\frac{4}{3})=0$				
			4	$2-(-\frac{3}{4})(-1)=1.25$	2	−1	0	0
					0	1.5	−1	0
					0	0	1.333	−1
					0	0	0	1.25

improved by careful programming. If the standard matrix facilities are unavailable or unsuitable, perhaps due to dimension restrictions, then it is not a difficult matter to program the operations in BASIC. Program 4.4 should be studied as an example of a program to add two rectangular matrices.

Program 4.4

```
10 PRINT "PROGRAM TO ADD TWO MATRICES, A = B + C"
20 PRINT "TYPE IN NUMBER OF ROWS"
30 INPUT N
40 PRINT "TYPE IN NUMBER OF COLUMNS"
50 INPUT M
60 DIM A(N,M),B(N,M),C(N,M)
70 PRINT "TYPE IN MATRIX B, ROW-BY-ROW, ELEMENTS SEPARATED BY 'RETURN'"
80 FOR I = 1 TO N
90 FOR J = 1 TO M
100 INPUT B(I,J)
110 NEXT J
115 PRINT "END OF ROW";I
120 NEXT I
130 PRINT "TYPE IN MATRIX C, ROW-BY-ROW, ELEMENTS SEPARATED BY 'RETURN'"
140 FOR I = 1 TO N
150 FOR J = 1 TO M
160 INPUT C(I,J)
170 NEXT J
175 PRINT "END OF ROW";I
180 NEXT I
190 FOR I = 1 TO N
200 FOR J = 1 TO M
210 A(I,J) = B(I,J) + C(I,J)
220 NEXT J
230 NEXT I
240 PRINT"MATRIX A NOW LISTED, COLUMN-BY-COLUMN"
250 FOR J = 1 TO M
260 PRINT"COLUMN";J;"OF MATRIX A"
270 FOR I = 1 TO N
280 PRINT A(I,J)
290 NEXT I
300 PRINT "PRESS ANY KEY TO CONTINUE"
310 GETA$: IF A$ ="" THEN 310
320 NEXT J
330 END
```

4.6 Introduction to Sorting

'Sorting' is the process of rearranging data according to specified criteria. The microcomputer is a very useful device for carrying out this operation and there are numerous examples in civil engineering where data have to be processed. As an illustration of a typical sorting procedure, we consider the problem of arranging a given sequence of numbers into descending numerical order. We examine two principles which can form a basis for carrying out this procedure.

4.6.1 Sorting by Successive Selection of the Maximum Number

Imagine that the given sequence is located in a one-dimensional array, $A(1), A(2), A(3), \ldots A(N)$, and that there is another one-dimensional array, $B(1), B(2), B(3), \ldots B(N)$, initially containing zeros. We scan array A to locate the maximum number, say $A(I)$, transfer $A(I)$ to $B(1)$ and place zero in $A(I)$. The process is repeated to locate the maximum number in the revised array A and this is transferred to $B(2)$. The process is continued until all the A numbers have been

Program 4.5

```
10 REM THIS PROGRAM SORTS A SET OF NUMBERS INTO DESCENDING ORDER BY
20 REM SUCCESSIVE SELECTION OF THE MAXIMUM NUMBER
25 PRINT "TYPE IN NUMBER OF NUMBERS N":INPUT N
30 DIM A(N),B(N)
40 PRINT"TYPE IN NUMBERS, ONE PER LINE"
45 FOR I = 1 TO N: INPUT A(I): NEXT
50 FOR I = 1 TO N
60 J = 1
70 K = 2
80 X = A(J) - A(K)
90 IF X > 0 THEN 130
100 J = K
110 IF J = N THEN 160
120 GOTO 140
130 IF K = N THEN 160
140 K = K + 1
150 GOTO 80
160 B(I)=A(J)
170 A(J)= -10E10
180 NEXT I
190 PRINT "SORTING COMPLETE, SEQUENCE NOW LISTED"
200 FOR I = 1 TO N
210 PRINT "B";I;"=";B(I)
220 PRINT "PRESS ANY KEY TO CONTINUE"
230 GETA$: IF A$="" THEN 230
240 NEXT I
250 PRINT "LISTING COMPLETE"
260 END
```

transferred into *B*. Program 4.5 performs this operation but care is needed if negative numbers can occur. If only positive numbers occur then line 170 would be written $A(J) = 0$, however, this means that if a negative number is encountered and compared with any of the 'planted' zeros, a zero will be chosen as numerically greater. If it is desired to handle lists which might contain negative numbers, then a 'large' negative number should be used in line 170.

4.6.2 Sorting by Successive Interchange of Adjacent Pairs of Numbers

In this procedure, sometimes called 'bubble sort', the first pair of numbers are compared and if the second number is larger than the first, then the numbers are exchanged. Succeeding pairs of numbers are examined in the same way and exchanged or not exchanged as the case may be. A record of the position of the *last* exchange is noted. The process is repeated, up to the position just before the last exchange, until no further exchanges are recorded. The sequence of numbers is then in descending numerical order. Program 4.6 carries out this process.

Exercises

4.1 If A is a (3 × 4) matrix and B is a (4 × 3) matrix, write a program to
   (i) input A and B
   (ii) form B^T

Program 4.6

```
READY.

 10 REM BUBBLE SORT PROGRAM
 20 PRINT "TYPE IN NUMBER OF NUMBERS TO BE SORTED"
 30 INPUT N
 40 DIM A(N)
 50 PRINT "TYPE IN NUMBERS, ONE PER LINE"
 60 FOR I = 1 TO N: INPUT A(I): NEXT I
 65 M=N: C=1
 70 GOSUB 300
 80 PRINT "SORTING COMPLETE, SEQUENCE NOW LISTED"
 90 FOR I = 1 TO N
 100 PRINT "A";I;"=";A(I)
 110 PRINT "PRESS ANY KEY TO CONTINUE"
 120 GETA$: IF A$="" THEN 120
 130 NEXT I
 140 PRINT "LISTING COMPLETE"
 150 END
 300 FOR I = 1 TO M-1
 310 IF A(I)>= A(I+1) THEN 360
 320 P = A(I)
 330 A(I)=A(I+1)
 340 A(I+1) = P
 350 C = I
 360 NEXT I
 365 IF M = C GOTO 400
 370 M = C
 380 B = C-1
 390 IF B > 0 THEN 300
 400 RETURN
```

(iii) form and print $(A + B^T)$

4.2 Repeat Exercise 4.1 with A an $(N \times M)$ matrix and B an $(M \times N)$ matrix.

4.3 A is an $(N \times N)$ matrix. Write a program to compute and print the sums of the squares of the elements in each row of A.

4.4 Write a program to

(a) input a one-dimensional array of N numbers;

(b) scan the array, record and print:

(i) the number of zeros (O)

(ii) the number of positive numbers (P)

(iii) the number of negative numbers (Q)

(c) perform the check $O + P + Q = N$ and display the result.

4.5 Write a program to:

(a) input an $(N \times N)$ array

(b) scan the array to locate the maximum number

(c) print the maximum number and its address in the array.

4.6 Given a one-dimensional array of N positive numbers, write a program to list the numbers in ascending order.

Further Reading

Broyden, C. G. *Basic Matrices*. MacMillan, 1975.

Flores, I. *Computer Sorting*. Prentice-Hall, 1969.

Steinberg, D. I. *Computational Matrix Algebra*. McGraw-Hill, 1974.

5

Numerical Methods

5.1 Introduction

The increasing availability of computers has been accompanied by a thorough review of the traditional methods used for numerical analysis in civil engineering. New methods have evolved, influenced by the power and speed of the computer, and old methods have sometimes been discarded and sometimes reformulated in more computer-orientated styles. In this chapter we examine some of the more important numerical methods which are of general use in civil engineering.

5.2 Numerical Integration by Trapezoidal and Simpson's Rules

Suppose we have a function

$$y = f(x) \tag{5.1}$$

which can be represented as in Fig. 5.1a and for which we have a sequence of numerical values of y; $h_1, h_2, h_3, \ldots$ at equal intervals a along the x axis. The area between the curve and the x axis is obtained approximately by adding the areas of the trapezia formed by drawing chords between the ordinates, i.e.

$$\text{Area} = \frac{a}{2}(h_1 + h_2) + \frac{a}{2}(h_2 + h_3) + \ldots + \frac{a}{2}(h_{n-1} + h_n)$$

$$= \frac{a}{2}\left(h_1 + 2\sum_{p=2}^{p=n-1} h_p + h_n\right) \tag{5.2}$$

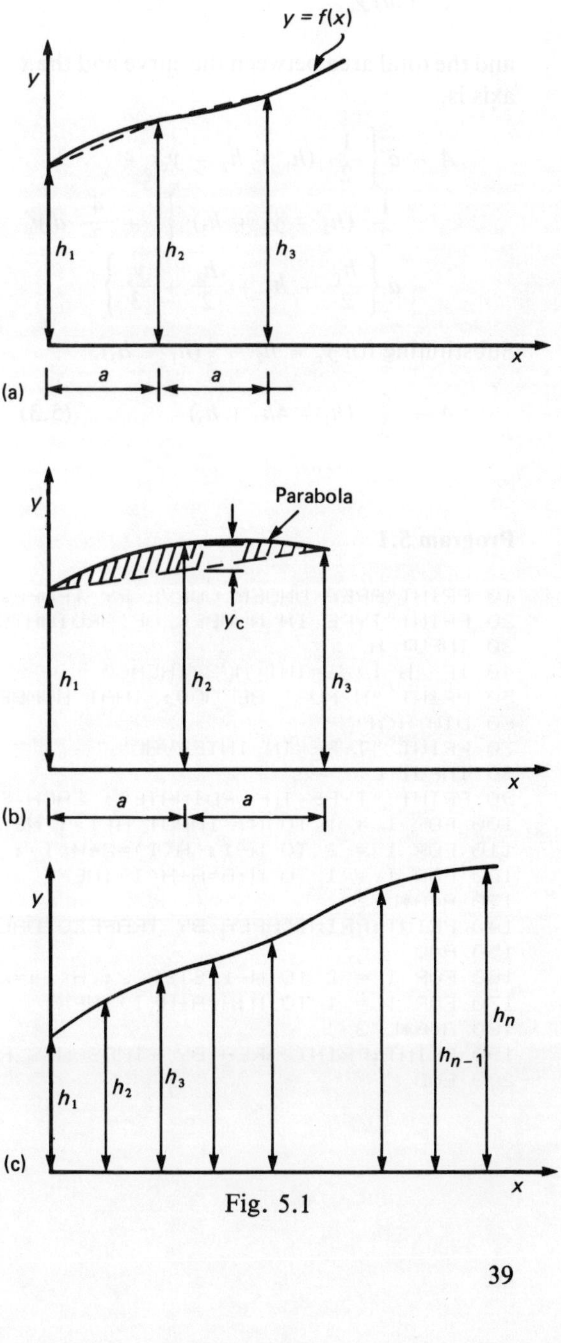

Fig. 5.1

Thus the function (curve) is replaced by straight lines between each successive pair of ordinates. This is the 'trapezoidal' rule. A more accurate evaluation of area may generally be obtained using Simpson's rule. In this rule, the function is replaced by a parabolic segment between successive groups of three ordinates as shown in Fig. 5.1b. The shaded area in Fig. 5.1b is, therefore,

$$\frac{2}{3}\ (2a)\, y_c$$

and the total area between the curve and the x axis is,

$$A = a\left\{\frac{1}{2}\ (h_1 + h_2 - y_c) + \frac{1}{2}\ (h_2 - y_c + h_3)\right\} + \frac{4}{3}\ a\, y_c$$

$$= a\left\{\frac{h_1}{2} + h_2 + \frac{h_3}{2} + \frac{y_c}{3}\right\}$$

Substituting for $y_c = h_2 - \frac{1}{2}(h_1 + h_3)$,

$$A = \frac{a}{3}\ (h_1 + 4h_2 + h_3) \qquad (5.3)$$

Hence in the general case, Fig. 5.1c, with n ordinates ($n - 1$ intervals) the total area is

$$A = \frac{a}{3}\ \{(h_1 + 4h_2 + h_3) + (h_3 + 4h_4 + h_5) + \ldots\}$$

$$A = \frac{a}{3}\ \{\text{sum of end ordinates} + 4(\text{sum of even ordinates}) + 2(\text{sum of other odd ordinates})\} \qquad (5.4)$$

Simpson's rule requires that the area be divided into an *even* number of intervals, i.e. an *odd* number of ordinates. Program 5.1 will compute the area under a curve, represented by a series of ordinates, by both the trapezoidal and Simpson's rules. The number of ordinates N is a program parameter. That is, N will be input each time the program is run. Program lines 40 and 50 serve to check that N is odd as required in the Simpson's rule application.

Program 5.1

```
10 PRINT"AREA UNDER CURVE BY TRAPEZOIDAL AND SIMPSON'S RULES"
20 PRINT"TYPE IN NUMBER OF ORDINATES"
30 INPUT N
40 IF (N-1)/2=INT(N/2)THEN60
50 PRINT "N MUST BE ODD; THAT NUMBER WAS EVEN":GOTO 20
60 DIM H(N)
70 PRINT "TYPE IN INTERVAL"
80 INPUT L
90 PRINT "TYPE IN ORDINATES; EACH FOLLOWED BY 'RETURN'"
100 FOR I = 1 TO N: INPUT H(I): NEXT
110 FOR I = 2 TO N-1: H(I)=2*H(I): NEXT
120 FOR I = 1 TO N:A=A+H(I):NEXT
130 A=A*L/2
140 PRINT:PRINT"AREA BY TRAPEZOIDAL RULE = ";A
150 A=0
160 FOR I = 2 TO N-1 STEP 2: H(I)=2*H(I):NEXT
170 FOR I = 1 TO N:A=A+H(I):NEXT
180 A=A*L/3
190 PRINT:PRINT"AREA BY SIMPSON'S RULE =";A
200 END
```

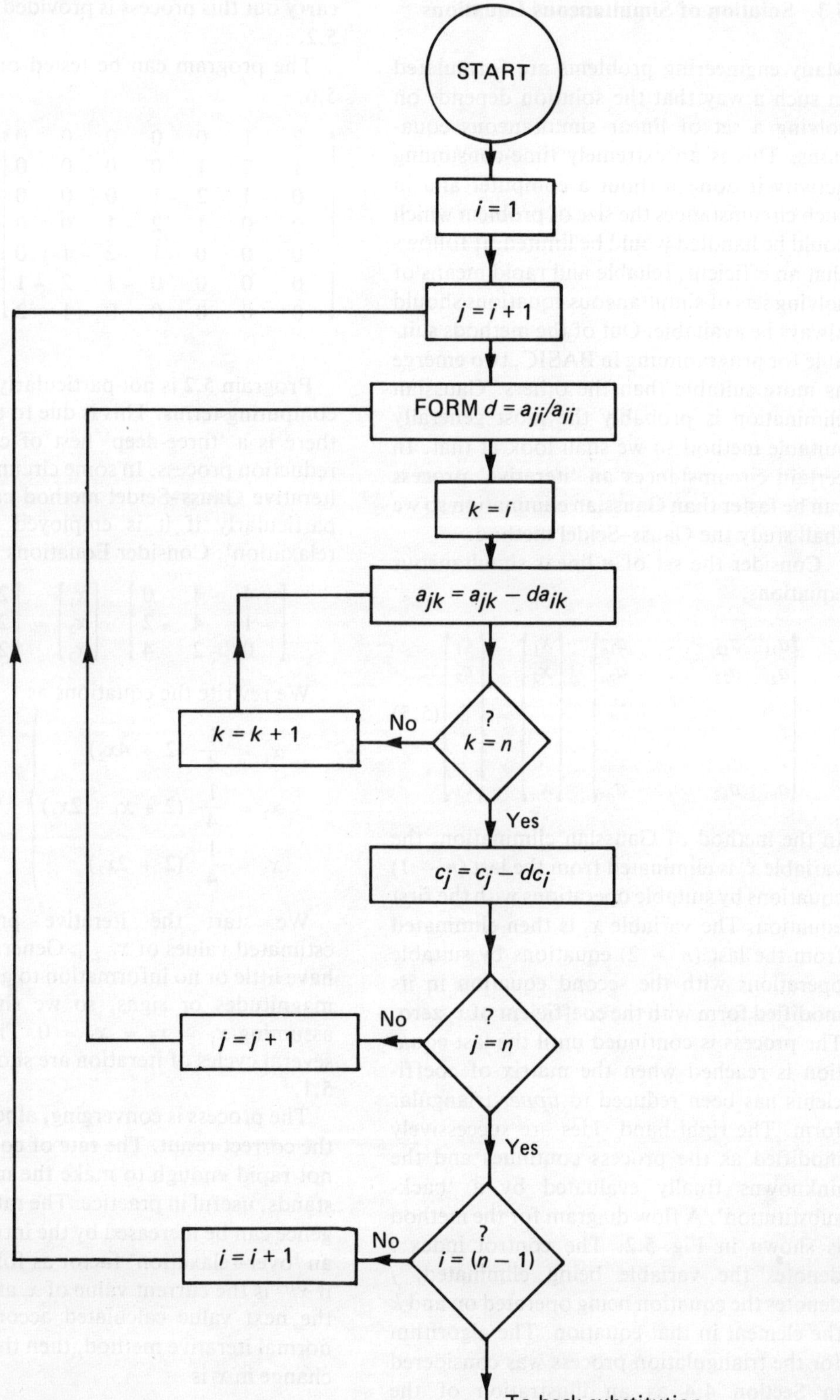

Fig. 5.2 Flow diagram for Gaussian elimination.

5.3 Solution of Simultaneous Equations

Many engineering problems are formulated in such a way that the solution depends on solving a set of linear simultaneous equations. This is an extremely time-consuming activity if done without a computer and in such circumstances the size of problem which could be handled would be limited. It follows that an efficient, reliable and rapid means of solving sets of simultaneous equations should always be available. Out of the methods suitable for programming in BASIC, two emerge as more suitable than the others. Gaussian elimination is probably the most generally suitable method so we shall look at that. In certain circumstances an 'iterative' process can be faster than Gaussian elimination so we shall study the Gauss–Seidel method.

Consider the set of n linear simultaneous equations,

$$\begin{bmatrix} a_{11} & a_{12} & \cdots & a_{1n} \\ a_{21} & a_{22} & \cdots & a_{2n} \\ \cdot & & & \cdot \\ \cdot & & & \cdot \\ \cdot & & & \cdot \\ a_{n1} & a_{n2} & \cdots & a_{nn} \end{bmatrix} \begin{bmatrix} x_1 \\ x_2 \\ \cdot \\ \cdot \\ \cdot \\ x_n \end{bmatrix} = \begin{bmatrix} c_1 \\ c_2 \\ \cdot \\ \cdot \\ \cdot \\ c_n \end{bmatrix} \quad (5.5)$$

In the method of Gaussian elimination, the variable x_1 is eliminated from the last $(n - 1)$ equations by suitable operations with the first equation. The variable x_2 is then eliminated from the last $(n - 2)$ equations by suitable operations with the second equation in its modified form with the coefficient of x_1 zero. The process is continued until the last equation is reached when the matrix of coefficients has been reduced to *upper* triangular form. The right-hand sides are successively modified as the process continues and the unknowns finally evaluated by a 'back-substitution'. A flow diagram for the method is shown in Fig. 5.2. The control index i denotes the variable being eliminated, j denotes the equation being operated on and k the element in that equation. The algorithm for the triangulation process was considered in Section 4.4 as an illustration of the approach to 'indexing'. A BASIC program to carry out this process is provided in Program 5.2.

The program can be tested on Equations 5.6.

$$\begin{bmatrix} 2 & -1 & 0 & 0 & 0 & 0 & 0 \\ -1 & 2 & -1 & 0 & 0 & 0 & 0 \\ 0 & -1 & 2 & -1 & 0 & 0 & 0 \\ 0 & 0 & -1 & 2 & -1 & 0 & 0 \\ 0 & 0 & 0 & -1 & 2 & -1 & 0 \\ 0 & 0 & 0 & 0 & -1 & 2 & -1 \\ 0 & 0 & 0 & 0 & 0 & -1 & 2 \end{bmatrix} \begin{bmatrix} 80 \\ 150 \\ 200 \\ 220 \\ 200 \\ 150 \\ 80 \end{bmatrix} = \begin{bmatrix} 10 \\ 20 \\ 30 \\ 40 \\ 30 \\ 20 \\ 10 \end{bmatrix} \quad (5.6)$$

Program 5.2 is not particularly efficient in computing terms. This is due to the fact that there is a 'three-deep' nest of cycles in the reduction process. In some circumstances the iterative Gauss–Seidel method can be faster particularly if it is employed with 'over-relaxation'. Consider Equations 5.7

$$\begin{bmatrix} 4 & -4 & 0 \\ -1 & 4 & -2 \\ 0 & -2 & 4 \end{bmatrix} \begin{bmatrix} x_1 \\ x_2 \\ x_3 \end{bmatrix} = \begin{bmatrix} 2 \\ 2 \\ 2 \end{bmatrix} \quad (5.7)$$

We rewrite the equations as

$$\left.\begin{aligned} x_1 &= \frac{1}{4}\,(2 + 4x_2) \\ x_2 &= \frac{1}{4}\,(2 + x_1 + 2x_3) \\ x_3 &= \frac{1}{4}\,(2 + 2x_2) \end{aligned}\right\} \quad (5.8)$$

We start the iterative process with estimated values of $x_{1,\,2,\,3}$. Generally we shall have little or no information to guide us as to magnitudes or signs, so we shall start by assuming $x_1 = x_2 = x_3 = 0$. The results of several cycles of iteration are shown in Table 5.1.

The process is converging, albeit slowly, to the correct result. The rate of convergence is not rapid enough to make the method, as it stands, useful in practice. The rate of convergence can be increased by the introduction of an 'over-relaxation' factor as follows: if $x_i^{(r)}$ is the current value of x_i and if $\bar{x}_i^{(r+1)}$ is the next value calculated according to the normal iterative method, then the 'apparent' change in x_i is

$$(\bar{x}_i^{(r+1)} - x_i^{(r)})$$

Program 5.2

```
10 PRINT "SOLUTION OF SIMULTANEOUS EQUATIONS BY GAUSSIAN ELIMINATION"
20 PRINT:PRINT"TYPE IN NUMBER OF EQUATIONS":INPUT N
30 DIM A(N,N),X(N),B(N)
40 PRINT"TYPE IN MATRIX A"
50 FOR I = 1 TO N:PRINT
60 PRINT"TYPE ROW";I;"ONE NUMBER PER LINE"
70 FOR J = 1 TO N
80 INPUT A(I,J)
90 NEXT J
100 NEXT I
110 PRINT"TYPE IN RIGHT-HAND SIDES"
120 FOR I = 1 TO N
130 INPUT B(I)
140 NEXT I
145 PRINT"COMPUTING, PLEASE WAIT"
150 FOR I = 1 TO N-1
160 FOR J = I+1 TO N
170 D=A(J,I)/A(I,I)
180 FOR K = 1 TO N
190 A(J,K)=A(J,K)-D*A(I,K)
200 NEXT K
210 B(J)=B(J)-D*B(I)
220 NEXT J
230 NEXT I
240 X(N)=B(N)/A(N,N)
250 FOR I =1 TO N-1
260 J=N-I
270 D=0
280 FOR K = J+1 TO N
290 D=D+A(J,K)*X(K)
300 NEXT K
310 X(J)=B(J)/A(J,J)-D/A(J,J)
320 NEXT I
330 PRINT"VALUES OF X NOW LISTED"
340 FOR I = 1 TO N
350 PRINT"X";I;"=";X(I)
352 PRINT"PRESS ANY KEY TO CONTINUE"
355 GET A$:IF A$ = "" THEN 355 ELSE 370
370 NEXT I
380 END
```

Now an 'over-relaxation' factor ω is applied to the change so that the next value of x_i is taken to be

$$x_i^{(r+1)} = x_i^{(r)} + \omega\,(\bar{x}_i^{(r+1)} - x_i^{(r)})$$

The optimum value of ω lies between 1 and 2, depends on the circumstances and can only be determined during the process. A value of ω of 1.5 or 1.6 will usually give an acceptable

Table 5.1 Gauss–Seidel Iteration

Cycle / Variable	1	2	3	4	5	6	7	Exact solution
x_1	0	0.5	1.125	1.688	1.97	2.11	2.18	2.25
x_2	0	0.625	1.1875	1.469	1.61	1.68	1.72	1.75
x_3	0	0.8125	1.0938	1.234	1.30	1.34	1.36	1.375

rate of convergence. Normally a Gauss–Seidel iteration program would have to include a convergence test based on a predetermined criterion. This is rather difficult to program and in the interactive mode can be avoided by listing the values of the variables after a fixed number of cycles and asking the operator to input a revised value of the over-relaxation factor if further iteration is required. This is a good example of the power of the interactive mode of computing. Program 5.3 will perform a Gauss–Seidel iteration with 10 cycles of iteration between each input of a fresh over-relaxation factor.

Example 5.1

Use the data from Equation 5.6 with the Gauss–Seidel program to determine the optimum value of the over-relaxation factor ω for these equations. The results are shown in Fig. 5.3 from which it is seen that the optimum value of ω is of the order of 1.5.

Most civil engineering systems are well defined physically and when the theoretical formulation leads to a set of simultaneous equations these will usually be stable, well behaved and will give reliable results. Under certain circumstances this is not so and 'ill-conditioned' equations result. This usually means that the mathematical description of the physical system is sensitive to small changes in the data. A coefficient matrix in which the diagonal elements are significantly greater than the off-diagonal elements (diagonally dominant) will generally be 'well conditioned'. The use of 'double length' arithmetic (twice the number of significant figures) can help in these circumstances but it is better

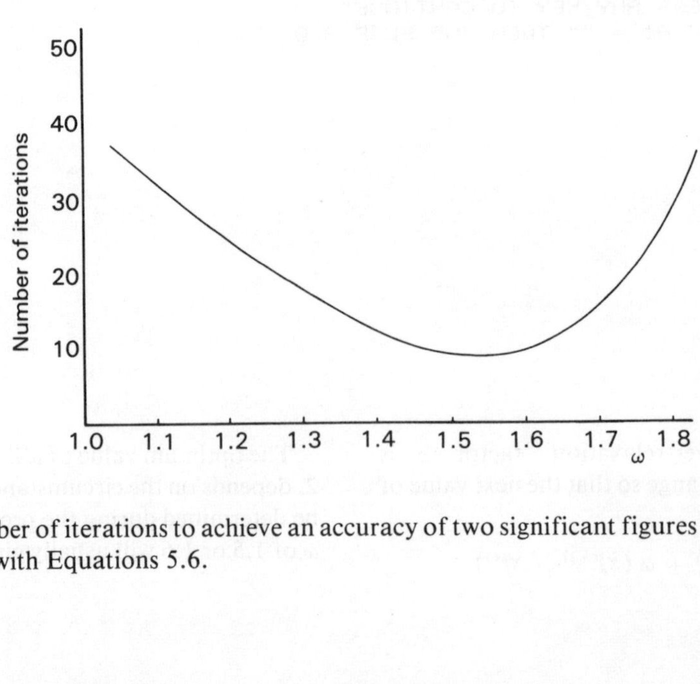

Fig. 5.3 Number of iterations to achieve an accuracy of two significant figures using the Gauss–Seidel method with Equations 5.6.

Program 5.3

```
10 PRINT"SOLUTION OF SIMULTANEOUS EQUATIONS BY GAUSS-SEIDEL ITERATION"
20 PRINT"EQUATIONS ARE AX=B; OVER-RELAXATION FACTOR = W"
30 PRINT"TYPE IN NUMBER OF EQUATIONS N":INPUT N
40 DIM A(N,N),X(N),B(N)
50 FOR I=1 TO N:PRINT
60 PRINT"ENTER ROW";I;"EACH NUMBER FOLLOWED BY 'RETURN'"
70 FOR J=1 TO N
80 INPUT A(I,J)
140 NEXT J
150 NEXT I
160 PRINT"TYPE IN RIGHT-HAND SIDES B, ONE NUMBER PER LINE"
170 FOR I=1 TO N:INPUT B(I): NEXT
180 PRINT"TYPE IN O R FACTOR W": INPUT W
185 PRINT"COMPUTING, PLEASE WAIT"
190 FOR K=1 TO 10
200 FOR I=1 TO N
210 A=0
220 FOR J=1 TO N
230 A=A+A(I,J)*X(J)
235 NEXT J
240 A=(B(I)-A+A(I,I)*X(I))/A(I,I)
250 X(I)=X(I)+(A-X(I))*W
260 NEXT I
270 NEXT K
280 FOR I=1 TO N
290 PRINT"X";I;"=";X(I)
300 PRINT"PRESS ANY KEY TO CONTINUE"
310 GETA$:IF A$="" THEN 310
320 NEXT I
330 PRINT:PRINT"IS ANOTHER ITERATION REQUIRED? Y/N"
340 GETA$:IF A$=""THEN340
350 IF A$="N"THEN 390
360 IF A$="Y"THEN 370
370 PRINT"TYPE IN VALUE OF O R FACTOR"
380 INPUT W:GOTO 185
390 END
```

to use an alternative mathematical formulation if one can be found.

5.4 Eigenvalue Problems

Eigenvalue-type problems occur in many areas of civil engineering, particularly in structural theory. Problems of vibrations and stability are good examples of physical systems described by eigenvalue theory. The mathematical expression of an eigenvalue problem is,

$$[\mathbf{A}]\{\mathbf{y}\} = \lambda\{\mathbf{y}\} \qquad (5.9)$$

where [**A**] is an $(n \times n)$ matrix, {**y**} is an $(n \times 1)$ column vector (eigenvector), and λ is the eigenvalue (or latent root).

A system of equations such as Equation 5.9 will have n eigenvalues and corresponding eigenvectors.

From Equation 5.9,

$$[\mathbf{A}^{-1}]\{\mathbf{y}\} = \frac{1}{\lambda}\{\mathbf{y}\} \tag{5.10}$$

An iterative procedure for finding λ and the corresponding eigenvector $\{\mathbf{y}\}$ will now be described. It can be shown that this procedure will lead to the *largest* value of λ.

The process is started with an *assumed* eigenvector $\{\mathbf{y}\}_0$, so

$$[\mathbf{A}^{-1}]\{\mathbf{y}\}_0 \simeq \frac{1}{\lambda}\{\mathbf{y}\}_0$$

putting $[\mathbf{A}^{-1}]\{\mathbf{y}\}_0 = \{\mathbf{y}\}_1$, (5.11)

$$\{\mathbf{y}\}_1 \simeq \frac{1}{\lambda}\{\mathbf{y}\}_0$$

hence

$$\lambda = \frac{\{\mathbf{y}\}_0}{\{\mathbf{y}\}_1} \tag{5.12}$$

The ratio $\{\mathbf{y}\}_0/\{\mathbf{y}\}_1$ in Equation 5.12 cannot be formed since each $\{\mathbf{y}\}$ is a column vector. One method of proceeding is to form the ratios of corresponding elements in the $\{\mathbf{y}\}$'s and take the largest ratio. Some methods express the ratio using the sums of the elements in the column vectors.

Example 5.2

Find the largest eigenvalue and corresponding eigenvector of the matrix,

$$[\mathbf{A}] = \begin{bmatrix} 2 & -1 & 0 \\ -1 & 2 & -1 \\ 0 & -1 & 2 \end{bmatrix}$$

Now,

$$[\mathbf{A}]^{-1} = \frac{1}{4}\begin{bmatrix} 3 & 2 & 1 \\ 2 & 4 & 2 \\ 1 & 2 & 3 \end{bmatrix}$$

Take

$$\{\mathbf{y}\}_0 = \begin{bmatrix} 1 \\ 1 \\ 1 \end{bmatrix}$$

then

$$\{\mathbf{y}\}_1 = [\mathbf{A}^{-1}]\{\mathbf{y}\}_0 = \begin{bmatrix} 1.5 \\ 2.0 \\ 1.5 \end{bmatrix}$$

At this stage, after one iteration, $\lambda = 0.6667$.

Now,

$$\{\mathbf{y}\}_2 = [\mathbf{A}^{-1}]\{\mathbf{y}\}_1 = \begin{bmatrix} 2.5 \\ 3.5 \\ 2.5 \end{bmatrix}$$

and

$$\lambda = \frac{1.5}{2.5} = 0.60$$

Now,

$$\{\mathbf{y}\}_3 = [\mathbf{A}^{-1}]\{\mathbf{y}\}_2 = \begin{bmatrix} 4.25 \\ 6.00 \\ 4.25 \end{bmatrix}$$

and $\lambda = 2.5/4.25 = 0.5882$.

$$\{\mathbf{y}\}_4 = [\mathbf{A}^{-1}]\{\mathbf{y}\}_3 = \begin{bmatrix} 7.25 \\ 10.25 \\ 7.25 \end{bmatrix}$$

and $\lambda = 4.25/7.25 = 0.5862$.

This process is easily programmed using a matrix multiplication subroutine and may be extended to the computation of the sub-dominant eigenvalues and eigenvectors.

5.5 Regression Analysis

This analysis is useful in finding the 'best fit' of a set of experimental observations to a mathematical relationship. If the 'best' straight line is sought, then it is *linear* regression analysis.

The independent variable is x and this is assumed to be error free. The dependent variable is y and this is the result of experimental observation and thus contains errors. The general straight line is (Fig. 5.4)

$$y = mx + c \tag{5.13}$$

where m is the slope of the line and c is the intercept on the Y axis.

Suppose we have n values of x and corresponding observations y_0. We use the prin-

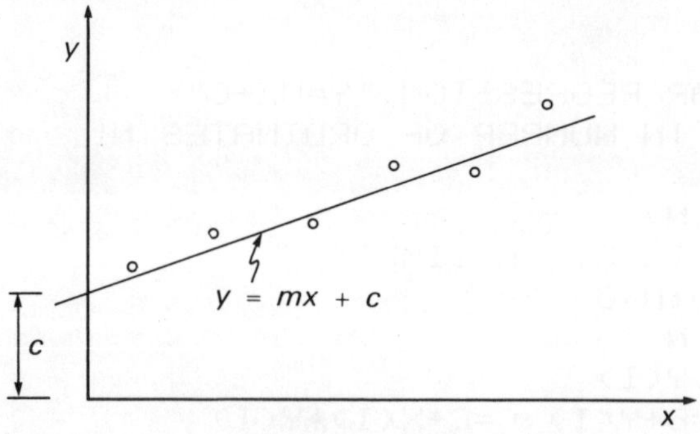

Fig. 5.4

ciple of 'least squares' to find the 'best' straight line to represent the relationship between y and x. The principle states that the sum of the squares of the 'errors' or 'deviations' should be a minimum, i.e. $\Sigma (y_0 - y)^2$ should be a minimum. Now,

$$\Sigma (y_0 - y)^2 = \Sigma (y_0 - mx - c)^2 \qquad (5.14)$$

Differentiating the right-hand side of Equation 5.14 partially with respect to m and c we obtain

$$c\Sigma x + m\Sigma x^2 = \Sigma x y_0 \qquad (5.15)$$

and

$$nc + m\Sigma x = \Sigma y_0 \qquad (5.16)$$

in which the summations are carried out from 1 to n. Solving Equations 5.15 and 5.16 we obtain

$$m = \frac{n\Sigma x y_0 - \Sigma y_0 \Sigma x}{n\Sigma x^2 - (\Sigma x)^2} \qquad (5.17)$$

and

$$c = \frac{\Sigma x^2 \Sigma y_0 - \Sigma x \Sigma x y_0}{n\Sigma x^2 - (\Sigma x)^2} \qquad (5.18)$$

The line $y = mx + c$ is the 'line of regression of y on x'.

Example 5.3

Table 5.2 gives the results of 12 laboratory observations y_0 at equal intervals of time t.

Find the line of regression of y on x. Now,

$$n = 12$$

$$\Sigma x = 660 \qquad \Sigma x^2 = 50\,600$$

$$\Sigma y_0 = 55.3 \qquad \Sigma x y_0 = 3468$$

Therefore

$$m = \frac{12 \times 3468 - 55.3 \times 660}{12 \times 50\,600 - 660^2} = 0.0298$$

and

$$c = \frac{50\,600 \times 55.3 - 660 \times 3468}{171\,600}$$

$$= 2.968$$

And thus the line is

$$y = 0.0298\,x + 2.968$$

A simple BASIC program to carry out this analysis is provided in Program 5.4.

The 'goodness of fit' of the straight line produced by linear regression needs to be assessed. Is the physical relationship represented by the data adequately described by a straight line? A means of testing this is by

Table 5.2

$t_{(x)}$	0	10	20	30	40	50	60	70	80	90	100	110
y_0	3	3.5	3.6	3.9	3.9	4.1	4.9	5.2	5.3	5.4	6.0	6.5

Program 5.4

```
10 PRINT"LINEAR REGRESSION, Y=MX+C"
20 PRINT"TYPE IN NUMBER OF ORDINATES N"
30 INPUT N
40 DIM X(N),Y(N)
50 PRINT"TYPE IN X,Y N TIMES"
55 A=0:B=0:C=0:D=0
60 FOR I=1 TO N
70 INPUT X(I),Y(I)
80 A=A+X(I):B=B+Y(I):C=C+X(I)*Y(I)
90 D=D+X(I)↑2
100 NEXT I
110 M=(N*C-A*B)/(N*D-A↑2)
120 C=(D*B-A*C)/(N*D-A↑2)
130 PRINT "LINE OF REGRESSION OF Y ON X IS":PRINT
140 PRINT"Y=";M;"X+";C
150 END
```

computing Pearson's correlation coefficient. If the experimental data lie close to a straight line, Pearson's coefficient will be near to ± 1. A coefficient near to zero indicates that a straight line is quite unsuitable to represent the relationship. The 'goodness of fit' of the straight line can be measured by calculating the standard deviation, a low value indicating a good fit.

Care is needed in using Equations 5.17 and 5.18 since in each case the denominator is obtained by subtraction. If the two quantities to be subtracted approach equality, then the accuracy of the computation is unreliable. In these circumstances an improved formulation of the equations should be used.

5.6 Polynomial Curve Fitting

There are many instances in civil engineering where it is useful to 'fit' a mathematical relationship, in the form of a polynomial, to experimental data. The general form of the polynomial is taken to be

$$y = a_0 + a_1x + a_2x^2 + a_3x^3 + \ldots + a_nx^n \qquad (5.19)$$

The largest power of x denotes the 'order' of the polynomial. Thus if $n = 2$ in Equation 5.19, we have a quadratic relationship between y and x:

$$y = a_0 + a_1x + a_2x^2 \qquad (5.20)$$

If it is assumed that the intervals x, at which observed values of y are available, are 'exact' then any errors are confined to the observations y. If an observed value y_i is known, then the corresponding 'residual' Δ_i is

$$\Delta_i = y_i - (a_0 + a_1x_i + a_2x_i^2 + \ldots + a_nx_i^n) \qquad (5.21)$$

The fitting of a polynomial involves choosing values for the coefficients $a_0, a_1, \ldots a_n$ which together minimize the sum of the squares of all the residuals as given by equations such as Equation 5.21. If there are p observed values of y and if E is the sum of the squares of the residuals then,

$$E = \sum_{i=1}^{p} \{y_i - (a_0 + a_1x_i + a_2x_i^2 + \ldots + a_nx_i^n)\}^2$$

$$= \sum_{i=1}^{p} \Delta_i^2 \qquad (5.22)$$

The procedure is to differentiate Equation 5.22 with respect to the coefficients $a_0, a_1, \ldots a_n$ in turn and equate to zero. The resulting equations are called the 'normal' equations and their solution gives the 'best' values of the coefficients in the polynomial of chosen order n. Suppose $n = 2$, then we have a quadratic relationship as in Equation 5.20 and,

$$\Delta_i = y_i - (a_0 + a_1x_i + a_2x_i^2) \qquad (5.23)$$

$$E = \sum_{i=1}^{p} \Delta_i^2 = \sum_{i=1}^{p} (y_i - a_0 - a_1x_i - a_2x_i^2)^2 \qquad (5.24)$$

Differentiating Equation 5.24 with respect to a_0, a_1 and a_2,

$$\frac{\partial E}{\partial a_0} = -2y_i + 2a_0 + 2a_1x_i + 2a_2x_i^2 = 0$$

$$\frac{\partial E}{\partial a_1} = -2x_iy_i + 2a_0x_i + 2a_1x_i^2 + 2a_2x_i^3 = 0$$

$$\frac{\partial E}{\partial a_2} = -2x_i^2y_i + 2a_0x_i^2 + 2a_1x_i^3 + 2a_2x_i^4 = 0$$

or

$$\begin{aligned} pa_0 + a_1\Sigma x_i + a_2\Sigma x_i^2 &= \Sigma y_i \\ a_0\Sigma x_i + a_1\Sigma x_i^2 + a_2\Sigma x_i^3 &= \Sigma x_iy_i \\ a_0\Sigma x_i^2 + a_1\Sigma x_i^3 + a_2\Sigma x_i^4 &= \Sigma x_i^2y_i \end{aligned} \qquad (5.25)$$

In each case the summation is taken over the p observations.

Equations 5.25 are the normal equations corresponding to a second-order polynomial. The normal equations in the general case of a polynomial of order n are:

$$\begin{bmatrix} p & \Sigma x_i & \Sigma x_i^2 & .. & \Sigma x_i^n \\ \Sigma x_i & \Sigma x_i^2 & \Sigma x_i^3 & .. & \Sigma x_i^{n+1} \\ . & & . & & . \\ . & & . & & . \\ \Sigma x_i^n & \Sigma x_i^{n+1} & \Sigma x_i^{n+2} & .. & \Sigma x_i^{2n} \end{bmatrix} \begin{bmatrix} a_0 \\ a_1 \\ . \\ . \\ a_n \end{bmatrix} = \begin{bmatrix} \Sigma y_i \\ \Sigma x_iy_i \\ . \\ . \\ \Sigma x_i^ny_i \end{bmatrix} \qquad (5.26)$$

Example 5.4

Fit a second degree polynomial to the following data

x	0	1	2	3	4	5
y	5	15	26	27	30	35

$$y = a_0 + a_1x + a_2x^2$$

$p = 6$, $n = 2$

$\Sigma x_i = 15$; $\Sigma x_i^2 = 55$; $\Sigma x_i^3 = 225$; $\Sigma x_i^4 = 979$

$\Sigma y_i = 138$; $\Sigma x_iy_i = 443$; $\Sigma x_i^2y_i = 1717$

The normal equations are, therefore,

$$\begin{bmatrix} 6 & 15 & 55 \\ 15 & 55 & 225 \\ 55 & 225 & 979 \end{bmatrix} \begin{bmatrix} a_0 \\ a_1 \\ a_2 \end{bmatrix} = \begin{bmatrix} 138 \\ 443 \\ 1717 \end{bmatrix}$$

from which,

$$a_0 = 5.61$$
$$a_1 = 10.7$$
$$a_2 = -1.02$$

Hence the polynomial is

$$y = 5.61 + 10.7x - 1.02x^2$$

5.7 Optimization of Constrained Linear Function

The ready availability of high-powered computers has encouraged engineering designers to seek 'optimum' or 'best' designs. The optimum design is usually defined as that design, described by a particular set of values of the design parameters, which minimizes an 'objective function'. The objective function is likely to be cost orientated but it may be weight orientated in certain circumstances.

Generally the minimization of the objective function will not be straightforward mathematically since the design parameters will be 'constrained' by certain requirements such as

$$\text{Stress} \leq \text{Limiting value} \tag{5.27}$$

and/or

$$\text{Deflection} \leq \text{Limiting value} \tag{5.28}$$

Some of the constraints will be *active* at the optimum design, i.e. the equality sign will apply in Equation 5.27, and/or Equation 5.28 and some will be *inactive*. We may, for example, produce an optimum design in which the deflection constraint was active and the deflection was thereby equal to the limiting value, while the stress constraint was inactive and the actual stress less than the limiting value.

If the objective function is linear in the design variables and if the constraints are also linear functions then, mathematically, the problem is one of 'linear programming'. Problems of this type can be solved graphically or analytically. The Simplex method is a well-known method for linear programming.

Engineering design often involves the selection of values of design parameters, depths of beams, cross-sectional areas of bars, second moments of inertia, for example from lists of 'discrete' values. The designer is not at liberty to choose *any* values but only those contained in the list. In the context of optimization, this is 'discrete variable programming' and special methods have been devised which are very suitable in these circumstances. One such method is called 'backtrack programming' and this will now be described in the context of a simple linear objective function of two variables, x and y, with any number of linear constraints.

We assume that the function to be minimized, F, can be expressed in the form,

$$F = Ax + By \tag{5.29}$$

and that x can be selected out of a list of M discrete values, and y out of a list of N values.

Any pair of values of x and y must conform to P constraints of the form:

$$\left.\begin{array}{l} C_1x + D_1y \geq H_1 \\ C_2x + D_2y \geq H_2 \\ \quad\cdot \qquad \cdot \qquad \cdot \\ \quad\cdot \qquad \cdot \qquad \cdot \\ C_px + D_py \geq H_p \end{array}\right\} \tag{5.30}$$

The problem is to find the combination of values of x and y which, while satisfying the constraints (Equation 5.30) minimizes the objective function F in Equation 5.29.

It is assumed that the sets of discrete values of x and y are stored in ascending order of magnitude so that x_M and y_N are the numerically greatest values in each set. The search commences by combining the *largest* value of x, x_M, with the *smallest* value of y, y_1. The constraints (Equation 5.30) are then applied to this pair of values of x and y. If one or more of the constraints is not satisfied, then the next, larger, value of y is chosen and the constraints applied again. This process is continued until a feasible combination is encountered. The value of the function F is then computed and stored. The index controlling the selection of x is then reduced by one and the process of scanning the y values repeated. Each feasible combination is recorded along with the corresponding value of the objective function. This type of search is 'exhaustive' in character and is guaranteed to locate the minimum value of the function.

Program 5.5 carries out a simple backtrack search through lists of discrete variables x and y subject to P linear constraints. The program can be tested on the data provided in Example 5.5.

Example 5.5

Using the following set of discrete values of x and y (706, 822, 919, 1140, 1250, 1406, 1550, 1792, 2290, 2737) obtain the minimum value of the function,

$$F = 6x + 8y$$

subject to

Program 5.5

```
10 PRINT"PROGRAM TO MINIMISE A FUNCTION OF TWO VARIABLES,F=AX+BY"
15 PRINT"BY BACKTRACK PROGRAMMING"
20 PRINT"TYPE IN COEFFICIENTS A,B"
30 INPUT A,B
40 PRINT"TYPE IN NUMBER OF X VALUES M, AND Y VALUES N IN FORM M,N"
50 INPUT M,N
60 DIM X(M),Y(N)
70 PRINT"TYPE IN";M;"VALUES OF X"
80 FOR I=1 TO M:INPUT X(I): NEXT I
90 PRINT"TYPE IN";N;"VALUES OF Y"
100 FOR I=1 TO N: INPUT Y(I): NEXT I
110 PRINT "TYPE IN NUMBER OF CONSTRAINTS P"
120 INPUT P
130 DIM C(P),D(P),H(P)
140 PRINT"TYPE IN";P;"VALUES OF C"
150 FOR I=1 TO P:INPUT C(I): NEXT I
160 PRINT"TYPE IN";P;"VALUES OF D"
170 FOR I=1 TO P:INPUT D(I):NEXT I
180 PRINT"TYPE IN";P;"VALUES OF H"
190 FOR I = 1 TO P:INPUT H(I):NEXT I
195 G=A*X(M) + B*Y(N)
197 PRINT"MAXIMUM VALUE OF THE FUNCTION F IS";G
200 FOR I = M TO 1 STEP -1
205 PRINT:PRINT"I=";I:PRINT
210 J=1
220 K=1
230 IF(C(K)*X(I)+D(K)*Y(J))<H(K) THEN 250
240 IF K = P THEN 260
245 K=K+1:GOTO 230
250 J=J+1: IF J>N THEN 315
252 GOTO 220
260 F=A*X(I)+B*Y(J)
270 PRINT"CURRENT VALUE OF X IS";X(I)
280 PRINT"CURRENT VALUE OF Y IS";Y(J)
290 PRINT"CURRENT VALUE OF FUNCTION F IS";F
300 PRINT"PRESS ANY KEY TO CONTINUE"
310 GETA$:IF A$=""THEN310
312 GOTO 320
315 PRINT"CONSTRAINTS NOT SATISFIED FOR REMAINING VALUES OF Y"
317 GOTO 330
320 NEXT I
330 END
```

$$
\begin{aligned}
x + y &\geq 2400\\
y &\geq 1200\\
x &\geq 750\\
2x + y &\geq 3500\\
x + 2y &\geq 3500
\end{aligned}
$$

Result: $F_{min} = 17\,500$ with $x = 1250$ and $y = 1250$.

Exercises

5.1 Improve Program 5.1 by introducing the following features:
  (i) odd ordinate check to apply only to Simpson's rule computation of area;
  (ii) build in an automatic repeat of the Simpson's rule computation with the given interval halved. Print out the result of this with a comment.

5.2 Adapt Program 5.2 to produce the inverse of matrix A.

5.3 Improve Program 5.3 by building in an automatic convergence test of a suitable kind.

5.4 Write a program to carry out the iterative 'largest eigenvalue' computation of Section 5.4. Use the data and results of Example 5.3 to test your program.

5.5 Adapt Program 5.4 to include the calculation of Pearson's correlation coefficient.

5.6 Write a program to fit a second degree polynomial to a given set of x,y data. Use the data from Example 5.4 to test your program.

Further Reading

Cope, A. J., Sawko, F. and Tickell, A. G. *Computer Methods for Civil Engineers.* McGraw-Hill (UK), 1982.

Gourlay, A. R. and Watson, G. A. *Computational Methods for Matrix Eigenproblems.* Wiley, 1973.

Hamming, R. W. *Introduction to Applied Numerical Analysis.* McGraw-Hill, 1971.

Kuo, S. S. *Computer Applications of Numerical Methods.* Addison-Wesley 1972.

Lee, J. D. and Lee, T. D. *Statistics and Computer Methods in BASIC.* Van Nostrand Reinhold, 1982.

Majid, K. I. *Introduction to Matrix and Numerical Methods for Civil Engineers.* Woodstock Publishing, 1980.

Seber, G. A. F. *Linear Regression Analysis.* Wiley, 1977.

6

Computer Output and Graphics

6.1 Introduction

Programs that are easy to use and understand, generally make effective use of the many display techniques available. A VDU enables displays to be produced which enhance programs and make them more interesting. Some of the displays can be produced by the use of the PRINT, TAB and SPC commands, however, many of the sophisticated graphics displays are produced by commands specific to one machine. The increasing availability of microcomputers has emphasized the importance of improving and extending control over screen displays. The use of the PRINT statement will be examined first and then the graphics capability of microcomputers discussed. Example programs will be given for Commodore microcomputers (4000 series) and Apple II computers as these machines cover the main features found on microcomputers.

6.2 Data Display Techniques

Data can be displayed either on a VDU or on paper by means of a line printer but one must remember that there are some important differences. For example, a screen is usually much smaller than paper and only a certain number of lines can be displayed at once. There are also special commands or keys that allow the cursor to be moved around a screen but this cannot be done on paper. The arrangement of data can be accomplished effectively by the PRINT statement and in some cases by the POKE statement. To use these statements on a microcomputer VDU we need to know how the screen is configured and how characters are displayed.

6.2.1 Screens and Memory Mapping

The majority of microcomputers use a technique known as 'memory mapping' to display characters on a screen. A specific area of RAM is set aside to store information that will be displayed on the screen. Each memory location stores an 8 bit coded byte. Special video circuitry scans the screen memory and by means of a character generator converts the code into a pattern of dots on the screen. As far as the programmer is concerned each memory location corresponds to a specific character location on the screen. The number of characters that can be displayed depends on the computer but varies from 23 characters per line to more than 80. The most popular is either 40 or 80, both of which are characteristic of Commodore and Apple microcomputers. Due to the variability of displays we will confine our discussion to Commodore and Apple II computers capable of displaying 40 characters per line.

Commodore microcomputers (4000 series) have 25 lines of 40 characters and require 1000 bytes of RAM memory to store the 1000 characters. The line and column numbers together with the screen memory addresses for the first character location of each line are shown in Fig. 6.1. An Apple II only displays 24 lines of 40 characters and requires a block of 1024 bytes of memory but not all of them are used (Fig. 6.2).

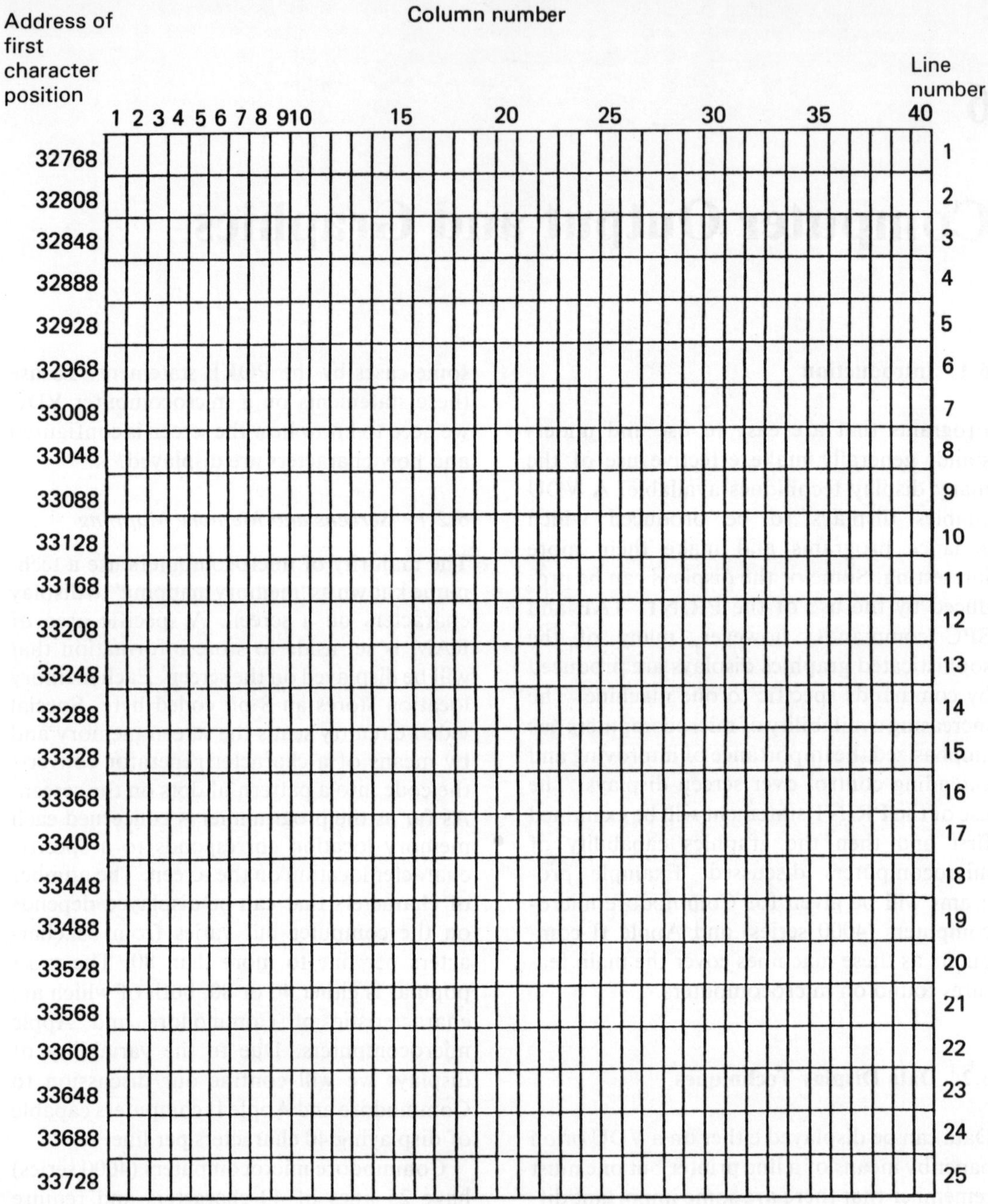

Fig. 6.1 Commodore 40-column screen.

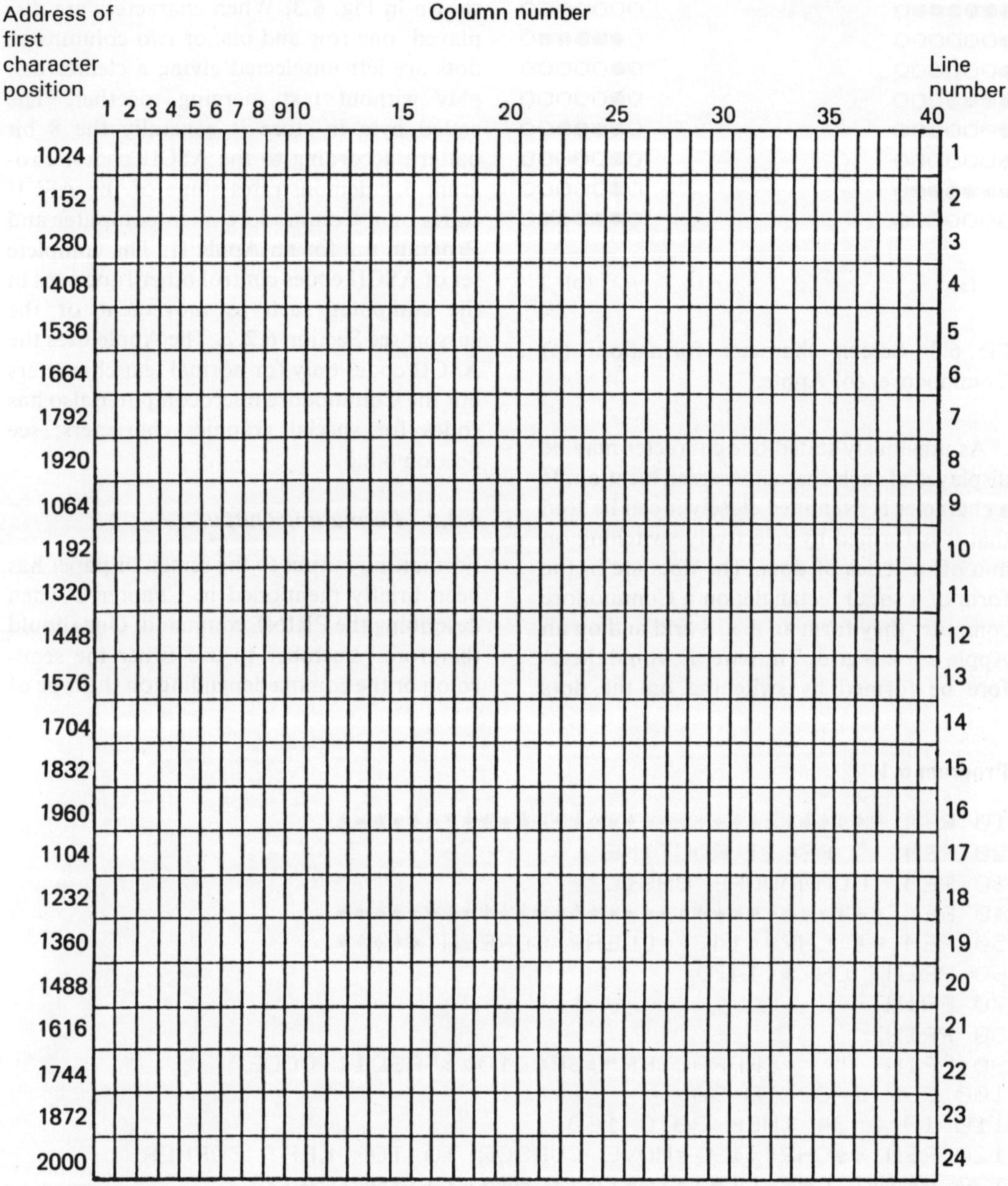

Fig. 6.2 Apple II 40-column screen.

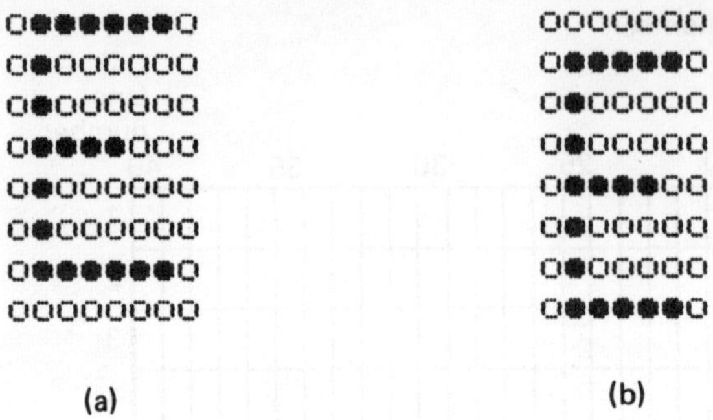

Fig. 6.3 Screen character formation: (a) Commodore, (b) Apple.

As previously stated one character may be displayed at each character screen location. If a character is examined closely it can be seen that it is formed by selectively switching on and off a series of dots. The dots are in the form of a small rectangle; on a Commodore computer they form an 8 × 8 grid and on an Apple a 7 × 8 grid. The letter E would therefore be formed by switching on the dots shown in Fig. 6.3. When characters are displayed, one row and one or two columns of dots are left unselected giving a clearer display without text merging together. The coded byte in store is generally the 8 bit pattern according to the ASCII codes. Program 6.1 demonstrates some of the ASCII codes on a Commodore microcomputer and Program 6.2 for an Apple II. The complete set of ASCII codes control other functions in the computer such as movement of the cursor; see Section 6.2.2. The Apple uses the ASCII codes only for normal text characters but the Commodore microcomputer also has codes for special graphics characters; see Section 6.2.5.

6.2.2 *Formatting Output*

Formatting output to the screen or paper has been briefly mentioned in Chapter 2 when describing the PRINT command. One should therefore remember to use either the semicolon or the comma depending on the type of

Program 6.1

```
10 REM ********************************
20 REM     ASCII ROUTINE
30 REM  COMMODORE BASIC
40 REM ********************************
50 REM ** CHR$(147)=CLEAR SCREEN *****
60 PRINT CHR$(147)
70 PRINT
80 PRINT
90 PRINT "   CHARACTER";SPC(10);"ASCII CODE"
100 FOR I=32 TO 90
110 IF I=34 THEN GOTO 160
120 REM **CHR$(19)=MOVE CURSOR TO TOP LEFT CORNER
130 PRINT CHR$(19);TAB(8);CHR$(I);TAB(24);I
140 FOR K=1 TO 300
150 NEXT K
160 NEXT I
170 FOR I=161 TO 223
180 PRINT CHR$(19);TAB(8);CHR$(I);TAB(24);I
190 FOR K=1 TO 300
200 NEXT K
210 NEXT I
220 END
```

Program 6.2

```
10 REM *********************************
20 REM    ASCII CODES
30 REM APPLESOFT BASIC
40 REM *********************************
50 REM *** HOME=CLEAR SCREEN
60 HOME
70 PRINT "    CHARACTER";SPC(10);"ASCII CODE"
80 FOR I=32 TO 90
90 VTAB 3
100 PRINT TAB(8);CHR$(I);TAB(24);I
110 FOR K=1 TO 300
120 NEXT K
130 NEXT I
140 END
```

spacing required. Output to a screen should be carefully planned using a grid such as shown in Fig. 6.1 or Fig. 6.2 so that data and instructions can be arranged clearly. There are some simple guidelines which should be followed to achieve this:

(a) Use headings to explain output.
(b) Align numerical data in columns.
(c) Avoid crowding too much information on to a screen.
(d) Ensure that the number of lines of data displayed is less than the maximum the screen can hold. Failing to do this could result in the initial lines being 'scrolled' off the top of the screen.
(e) Incorporate a *waiting* routine that enables users to read the information displayed before it is wiped out by further data.

Screen formatting on microcomputers is also aided by additional commands that enable the screen to be cleared and the printing position to be defined. For Apple II computers the following commands move the cursor and can be used in programs:

HOME	move cursor to top left-hand corner and clear screen
VTAB *n*	move cursor vertically to line *n* ($1 \leq n \leq 24$)
HTAB *n*	move cursor horizontally to column *n* ($1 \leq n \leq 40$)

With Commodore microcomputers there are special control keys which can also be used within a program. When used in a program these cursor controls are represented by graphic symbols as shown in Table 6.1. These controls can also be activated by printing their ASCII codes using the PRINT statement and the CHR$ function; see Section 6.2.4.

Program 6.3 demonstrates tabulation on a Commodore microcomputer, with all for-

Table 6.1 Cursor Control (Commodore)

Cursor control	Key	ASCII code
Clear screen	Reverse shifted S	147
Cursor down	Reverse Q	17
Cursor up	Reverse shifted Q	145
Cursor right	Reverse]	29
Cursor left	Reverse shifted]	157

Program 6.3 Listing and output

```
10 REM ********************************
20 REM FORMATTING OUTPUT
30 REM COMMODORE BASIC
40 REM ********************************
50 DATA 356,406,340,2980,4410
60 DATA 356,406,287,2690,3780
70 DATA 356,406,235,2200,3100
80 DATA 356,368,202,1890,2660
90 DATA 356,368,174,1630,2300
100 DATA 356,368,152,1420,2000
110 DATA 356,368,133,1290,1800
120 DATA 356,368,109,1050,1470
130 DATA 305,305,283,2480,3670
140 DATA 305,305,240,2240,3160
150 DATA 305,305,223,2090,2950
160 DATA 305,305,198,1850,2610
170 DATA 305,305,186,1740,2450
180 DATA 305,305,158,1480,2080
190 REM *** READ IN DATA ****
200 DIM DA(14,5)
210 FOR  I=1 TO 14
220 FOR J=1 TO 5
230 READ DA(I,J)
240 NEXT J
250 NEXT I
260 PRINT CHR$(147)
270 PRINT "END-BEARING STEEL H PILES"
280 PRINT
290 PRINT"MAXIMUM WORKING LOADS"
300 PRINT"==================================="
310 PRINT"SERIAL";TAB(11);":MASS";TAB(17);": MAX WORKING LOAD/PILE"
320 PRINT"SIZE";TAB(11);":";TAB(17);":GRADE 43A";TAB(28);":GRADE 50B"
330 PRINT"------------------------------------"
340 PRINT"MM X MM";TAB(11);":KG/M";TAB(17);":";TAB(21);"KN";
350 PRINT TAB(28);":";TAB(33);"KN"
360 PRINT"-------------------------------------"
370 FOR I=1 TO 14
380 PRINT DA(I,1);"X";DA(I,2);TAB(10);":";DA(I,3);
390 PRINT TAB(15);":";TAB(18);DA(I,4);TAB(28);":";TAB(31);DA(I,5)
400 NEXT I
410 END
```

Program 6.3 *(continued)*

```
END-BEARING STEEL H PILES

MAXIMUM WORKING LOADS
========================================
SERIAL       :MASS : MAX WORKING LOAD/PILE

SIZE         :     :GRADE 43A :GRADE 50B
----------------------------------------
MM X MM      :KG/M :   KN     :    KN
----------------------------------------
  356 X 406 : 340 : 2980      : 4410
  356 X 406 : 287 : 2690      : 3780
  356 X 406 : 235 : 2200      : 3100
  356 X 368 : 202 : 1890      : 2660
  356 X 368 : 174 : 1630      : 2300
  356 X 368 : 152 : 1420      : 2000
  356 X 368 : 133 : 1290      : 1800
  356 X 368 : 109 : 1050      : 1470
  305 X 305 : 283 : 2480      : 3670
  305 X 305 : 240 : 2240      : 3160
  305 X 305 : 223 : 2090      : 2950
  305 X 305 : 198 : 1850      : 2610
```

matting being achieved by the PRINT and TAB commands and through the use of the minus and equals sign.

6.2.3 *Menus and Selection*

Programs are often written to solve more than one problem depending on a variety of conditions. There may be several options from which a user can select one. The simplest way of selecting one option from many is to use a 'menu'. A menu displays the options and requests the user to select one by pressing a particular key. Program 6.4 demonstrates how a menu for a program that manipulates matrix routines can be displayed. The selection routine in lines 27Ø to 28Ø uses the GET statement.

6.2.4 *Simple Graphical Output*

The PRINT and TAB statements can also be used to provide simple graphical output on a screen or paper. An example of this is given in Program 6.5 and Fig. 6.4. The program prints to the screen a graph of the sine function between the angles 0 and 180 degrees. Line 7Ø prints a title with lines 8Ø to 11Ø outputting the horizontal axis and labelling the limits − 1 to 1. The graph itself is output in the FOR loop lines 12Ø to 14Ø. The print statement prints a character "I" as a vertical axis and the value of the sine is scaled to fit on the screen by the expression in the TAB function. The value is displayed by an "*" at the position defined in the TAB. With the Apple there are 40 character positions and a value between − 1 and 1 must be scaled. This is achieved by first noting the location of zero (i.e. character position 20) and adding this to the integer result of the sine value multiplied by 20. In this way negative and positive values of the sine function can be plotted.

6.2.5 *Poking to Screen Memory*

An alternative method of displaying charac-

Program 6.4

```
10 REM *******************************
20 REM     MENU SELECTION
30 REM APPLESOFT BASIC
40 REM *******************************
50 HOME
60 VTAB 2
70 HTAB 10
80 PRINT"MATRIX MENU SELECTION"
90 PRINT
100 PRINT
110 PRINT"   1.     ENTER MATRIX A"
120 PRINT
130 PRINT"   2.     ENTER MATRIX B"
140 PRINT
150 PRINT"   3.     INVERT MATRIX A"
160 PRINT
170 PRINT"   4.     INVERT MATRIX B"
180 PRINT
190 PRINT"   5.     MULTIPLY A BY B"
200 PRINT
210 PRINT"   6.     ADD A+B"
220 PRINT
230 PRINT"   7.     SUBTRACT A FROM B"
240 PRINT
250 PRINT
260 PRINT"          SELECT OPTION (1-7)"
270 GET A$
280 IF A$<"1" AND A$>"7" THEN 270
290 PRINT
300 PRINT
310 PRINT" OPTION ";A$;" SELECTED"
320 END
```

Program 6.5

```
10 REM ********************************
20 REM  GRAPH OF SINE FUNCTION
30 REM APPLESOFT BASIC
40 REM ********************************
50 HOME
60 DEF FNA(X)=SIN(X*3.142/180)
70 PRINT "      GRAPH OF SINE FUNCTION"
80 PRINT "-1";
90 HTAB 38
100 PRINT"1"
110 PRINT"----------------------------------------"
120 FOR I=0  TO 180 STEP 10
130 PRINT"I";TAB(20+(INT(FNA(I)*20)));"*"
140 NEXT I
150 END
```

Table 6.2 Apple II Character Set

ASCII code	Character
32	
33	!
34	"
35	#
36	$
37	%
38	&
39	'
40	(
41	)
42	*
43	+
44	,
45	-
46	.
47	/
48	0
49	1
50	2
51	3
52	4
53	5
54	6
55	7
56	8
57	9
58	:
59	;
60	<
61	=
62	>
63	?
64	@
65	A
66	B
67	C
68	D
69	E
70	F
71	G
72	H
73	I
74	J
75	K
76	L
77	M
78	N
79	O
80	P
81	Q
82	R
83	S
84	T
85	U
86	V
87	W
88	X
89	Y
90	Z
91	[
92	\
93	]
94	↑

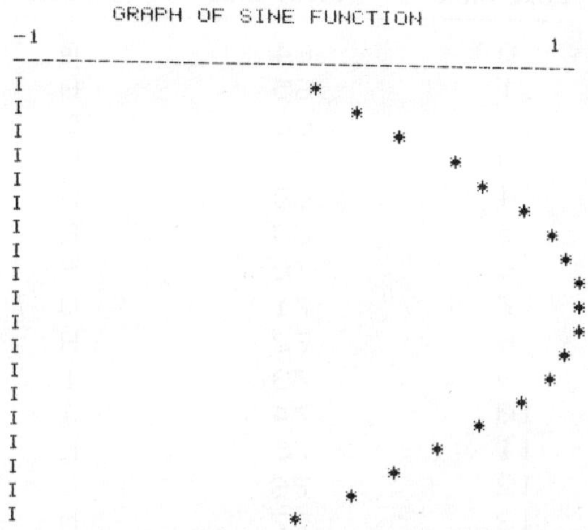

Fig. 6.4 Screen output of Program 6.5.

ters on a screen is by poking the codes directly into the screen memory locations. As the video circuitry scans this area anything poked into memory will be displayed. The format for the POKE command was outlined in Chapter 3 but we should consider the code required. This in fact depends on the computer being used. An Apple II microcomputer uses the standard ASCII code for the normal alphabetic, numeric and punctuation characters (Table 6.2). The Commodore microcomputers do not use standard ASCII codes to display characters on the screen. The reason for this is that Commodore machines have an extensive set of graphics characters allowing diagrams, graphs and text to be output on the same screen. The Commodore Poke codes and their associated characters are shown in Table 6.3. Alternatively the available characters can be demonstrated on a Commodore (4000 or 8000) microcomputer by running Program 6.6. The program displays all the possible characters by the FOR loops in lines 7Ø to 12Ø. In line 9Ø the cursor is controlled by printing ASCII codes to the screen. These cursor control codes are

Table 6.3 Commodore Character Set

Poke value	ASCII code	Character	Poke value	ASCII code	Character
0	64	@	48	112	0
1	65	A	49	113	1
2	66	B	50	114	2
3	67	C	51	115	3
4	68	D	52	116	4
5	69	E	53	117	5
6	70	F	54	118	6
7	71	G	55	119	7
8	72	H	56	120	8
9	73	I	57	121	9
10	74	J	58	122	:
11	75	K	59	123	;
12	76	L	60	124	<
13	77	M	61	125	=
14	78	N	62	126	>
15	79	O	63	127	?
16	80	P	64	128	
17	81	Q	97	161	▌
18	82	R	98	162	▄
19	83	S	99	163	▔
20	84	T	100	164	▁
21	85	U	101	165	▏
22	86	V	102	166	▒
23	87	W	103	167	▕
24	88	X	104	168	░
25	89	Y	105	169	◤
26	90	Z	106	170	▕
27	91	[	107	171	├
28	92	\	108	172	▗
29	93	]	109	173	└
30	94	↑	110	174	┐
31	95	←	111	175	▂
32	96		112	176	┌
33	97	!	113	177	┴
34	98	"	114	178	┬
35	99	#	115	179	┤
36	100	$	116	180	▎
37	101	%	117	181	▍
38	102	&	118	182	▐
39	103	'	119	183	▔
40	104	(	120	184	▀
41	105	)	121	185	▃
42	106	*	122	186	┘
43	107	+	123	187	▖
44	108	,	124	188	▝
45	109	−	125	189	┘
46	110	.	126	190	▘
47	111	/	127	191	▚

Poke value	ASCII code	Character	Poke value	ASCII code	Character
128			176		
129			177		
130			178		
131			179		
132			180		
133			181		
134			182		
135			183		
136			184		
137			185		
138			186		
139			187		
140			188		
141			189		
142			190		
143			191		
144			192		
145			225		
146			226		
147			227		
148			228		
149			229		
150			230		
151			231		
152			232		
153			233		
154			234		
155			235		
156			236		
157			237		
158			238		
159			239		
160			240		
161			241		
162			242		
163			243		
164			244		
165			245		
166			246		
167			247		
168			248		
169			249		
170			250		
171			251		
172			252		
173			253		
174			254		
175			255		

specific to Commodore machines and have the following meaning

147	Clear screen and move cursor to top left
19	move cursor to top left
17	move cursor down one line
29	move cursor right, one position
157	move cursor left, one position
145	move cursor up one line

Poking to the screen can be used to place borders round the screen to improve a display (Programs 6.7 and 6.8) or to produce a graphical output, see Section 6.3.4.

6.3 Graphics

The printing of a graph of the sine function in Program 6.5 is essentially a simple graphics routine. The values plotted were between -1 and $+1$ and hence each character position represented a value ± 0.05 horizontally and $\pm 10°$ vertically. The resolution on such a graph is adequate for demonstrations, however, we now consider how a higher resolution can be obtained.

6.3.1 Screen Resolution

On Commodore (4000 series) microcomputers and Apple II computers, text is placed on the screen with a resolution of 40 × 25 and 40 × 24 respectively. On a TRS-80 microcomputer it would be 64 × 16, and on Commodore's 8000 series 80 × 25. In Section 6.2.1 we noted that each character was made up of a series of dots, for example a Commodore character is 8 × 8 dots. Thus, if it were possible to switch on a single dot, the resolution which could be achieved would be 320 × 200. Unfortunately we would require 8 kilobytes of screen memory for this type of resolution. Because of the memory requirements, manufacturers have used a variety of methods to improve the resolution and yet keep memory requirements to a minimum. The Commodore 4000 series and the Apple II microcomputers can be used to demonstrate the methods by which manufacturers have improved screen resolution.

6.3.2 Text or Character Resolution

In this mode the resolution is based on single character positions. Although resolution is poor it is in fact the only mode possible on standard Commodore 4000 and 8000 series microcomputers. It can be used with comparative ease on these machines due to the fact that all screen locations from the top left-hand corner to the bottom right-hand corner are consecutive. This is not the case for Apple II computers and it is not recommended for these microcomputers.

This particular mode can be used to demonstrate how graphs are drawn and information poked to screen memory. Program 6.9 for a Commodore 4000 series microcomputer plots the liquid and plastic limit values of Table 3.1. The data are first read in and then scaled, lines 90 to 140 and 210 to 250. Scaling will produce column values between 1 and 40 for the liquid limit and row values between 1 and 25 for the plastic limit. Before a character is poked to the screen we need the address of the screen memory location. This is obtained using the following formula:

$$\text{Screen Memory Address} = 32768 + (\text{COL} - 1) + (40 * (\text{ROW} - 1))$$

Once the screen memory address has been calculated it should be checked to ensure that it is within screen memory limits as in lines 310 and 320. Failure to have such a check routine could allow a value to be poked into a vital area of memory resulting in the program crashing. Having done this the axes and labelling can be placed on the screen, lines 350 to 450. Axes should be drawn first just in case

Program 6.6

```
10 REM ******************************
20 REM     POKE AND PEEK VALUES
30 REM COMMODORE BASIC ONLY
40 REM ******************************
50 PRINT CHR$(147);
60 PRINT "       CHARACTER ";SPC(10);"POKE VALUE"
70 FOR I=0 TO 255
80 POKE 32860,I
90 PRINT CHR$(19);CHR$(17);CHR$(17);TAB(31);I
100 FOR K=1 TO 300
110 NEXT K
120 NEXT I
```

Program 6.7

```
10 REM ******************************
20 REM     BORDER ROUTINE
30 REM COMMODORE BASIC
40 REM ******************************
50 PRINT CHR$(147)
60 FOR I=32768 TO 32807
70 POKE I,102
80 POKE I+960,102
90 NEXT I
100 FOR I=32768 TO 33728 STEP 40
110 POKE I,102
120 POKE I+39,102
130 NEXT I
140 PRINT CHR$(19)
150 PRINT CHR$(17);CHR$(17)
160 PRINT TAB(10);"BORDER ROUTINE
170 GET A$
180 IF A$="" THEN GOTO 170
190 REM ** CHR$(17)= MOVE CURSOR DOWN ONE LINE
200 END
```

Program 6.8

```
10 REM ********************************
20 REM      BORDER ROUTINE
30 REM APPLESOFT BASIC(APPLE II)
40 REM ********************************
50 HOME
55 J=0
60 FOR I=1024 TO 2007 STEP 40
70 J=J+1
80 POKE I,42
90 POKE I+39,42
100 IF J=3 THEN I=I+8
110 IF J=3 THEN J=0
120 NEXT I
130 FOR I=1024 TO 1064
140 POKE I,42
150 POKE I+976,42
160 NEXT I
170 VTAB 3
180 HTAB 14
190 PRINT "BORDER ROUTINE"
200 GET A$
210 STOP
220 END
```

a point plots on one of them (Fig. 6.5).

Although the method gives poor results for graphs it can be used very effectively for bar charts.

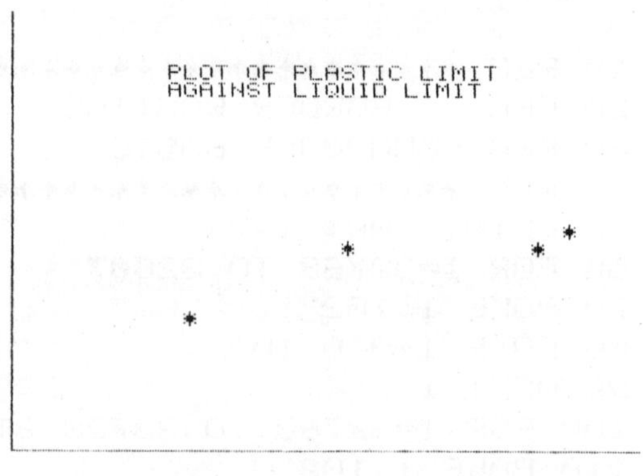

Fig. 6.5 Screen output of Program 6.9.

6.3.3 Low Resolution

'Low resolution' on microcomputers uses 'pixel' characters. These characters are switched on and off by special commands from BASIC. True pixel characters are found on TRS-80 microcomputers where each character location is subdivided into six blocks (Fig. 6.6). With this subdivision a TRS-80 has a resolution of 128 × 48. Special commands, for example,

SET *X*, *Y*

and

RESET *X*, *Y*

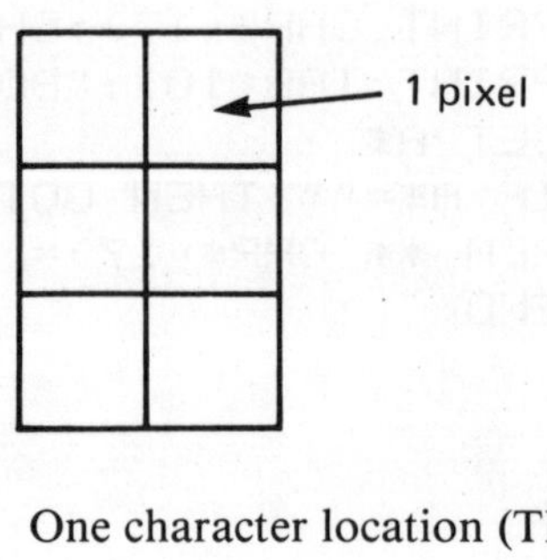

Fig. 6.6 One character location (TRS–80).

can be used to switch on or off any of the 6144 points that can be displayed on the screen.

A similar mode is available on Apple II computers by using the command:

GR

This command sets up a protected area of memory allowing a resolution of 40 × 48 points, in other words each character location is divided into two pixels as shown in Fig. 6.7.

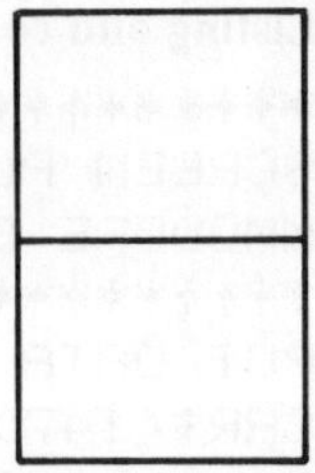

Fig. 6.7 Subdivision of Apple character location.

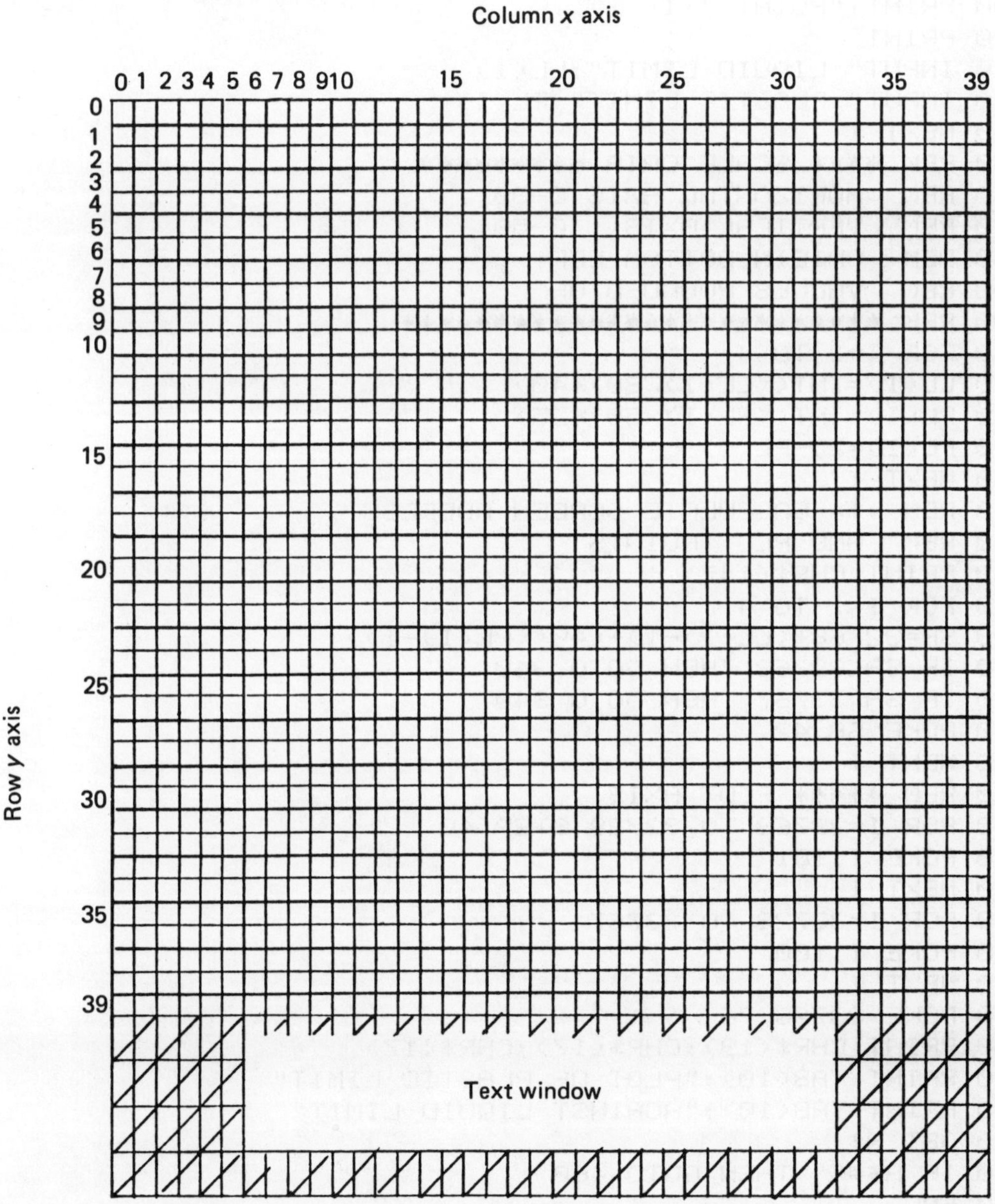

Fig. 6.8 Apple low-resolution graphics screen.

Program 6.9 Listing and copy of screen display

```
10 REM ********************************
20 REM   SCREEN PLOTTING LOW/RES
30 REM COMMODORE BASIC :4000 SERIES
40 REM ********************************
50 REM INPUT DATA
60 PRINT CHR$(147);CHR$(17);CHR$(17)
70 PRINT "         SCREEN PLOTTING"
80 PRINT " ENTER DATA"
90 FOR I=1 TO 4
100 PRINT "POINT ";I
110 PRINT
120 INPUT" LIQUID LIMIT";LL(I)
130 INPUT" PLASTIC LIMIT";PL(I)
140 NEXT I
150 REM **** SCALE DATA *********
160 REM  HORIZONTAL AXIS 0-80
170 REM  VERTICAL AXIS   0-50
180 REM  ORIGIN BOTTOM LEFT
190 REM  VALUES ROUNDED UP
200 REM *************************
210 FOR I=1 TO 4
220 LL(I)=INT((LL(I)/80)*40)
230 PL(I)=INT((PL(I)/50)*25)
240 PL(I)=25-PL(I)
250 NEXT I
260 REM *** CALCULATE SCREEN ADDRESS
270 REM  AND PLOT POINTS
280 PRINT CHR$(147)
290 FOR I=1 TO 4
300 SA=32768+(LL(I)-1)+(40*(PL(I)-1))
310 IF SA<32768 THEN GOTO 340
320 IF SA>33767 THEN GOTO 340
330 POKE SA,42
340 NEXT I
350 REM ***** PLOT AXIS
360 FOR I=32768 TO 33728 STEP 40
370 POKE I,101
380 NEXT I
390 FOR I=33728 TO 33767
400 POKE I,100
410 NEXT I
420 POKE 33728,76
430 PRINT CHR$(19);CHR$(17);CHR$(17)
440 PRINT TAB(10);"PLOT OF PLASTIC LIMIT"
450 PRINT TAB(10);"AGAINST LIQUID LIMIT"
460 GET A$
470 IF A$="" THEN GOTO 460
480 END
```

Although the resolution is 40 (0 to 39) horizontally and 48 (0 to 47) vertically, the last eight rows of low-resolution memory (four character rows) are set aside for text. This text window has the advantage that messages can be displayed on the screen without destroying low-resolution graphics (Fig. 6.8). The following functions are available for low-resolution plotting on the Apple II:

PLOT *Y*,*X*	Displays the pixel at row *Y*, column *X*.

Program 6.10

```
10 REM *********************************
20 REM SCREEN PLOTTING
30 REM LOW RESOLUTION
40 REM APPLESOFT BASIC : APPLE II
50 REM *********************************
60 DIM LL(4),PL(4)
70 HOME
80 VTAB 2
90 PRINT" ENTER DATA"
100 FOR I=1 TO 4
110 PRINT"POINT ";I
120 PRINT"    ENTER LL"
130 INPUT LL(I)
140 PRINT"    ENTER PL"
150 INPUT PL(I)
160 PRINT
170 NEXT I
180 REM *** SCALE DATA *****
190 REM HORIZONTAL AXIS 0-80
200 REM VERTICAL AXIS   0-30
210 REM TEXT WINDOW PRESERVED
220 REM ORIGIN BOTTOM LEFT
230 REM VALUES ROUNDED UP
240 FOR I=1 TO 4
250 LL(I)=INT((LL(I)/80)*40)
260 PL(I)=INT((PL(I)/30)*40)
270 PL(I)=40-PL(I)
280 NEXT I
290 REM *** PLOT AXIS ****
300 COLOR=12
310 HLIN 0,39 AT 39
320 VLIN 0,39 AT 0
330 REM *** PLOT DATA ****
340 FOR I=1 TO 4
350 PLOT LL(I),PL(I)
360 NEXT I
370 REM *** NOTE : NO LABELS CAN
380 REM BE PLOTTED ON THE LOW
390 REM RESOLUTION SCREEN
400 END
```

	Both X and Y can be numeric values, variables or expressions.
HLIN $X1$, $X2$ AT Y	Draws a horizontal line starting at column $X1$ and ending at column $X2$ in row Y
VLIN $Y1$, $Y2$ AT X	Draws a vertical line starting at row $Y1$ and ending at row $Y2$ in column X

If a colour monitor (VDU) is available then the plots and lines can be shown in different colours. A disadvantage of this mode on an Apple II is that text cannot be displayed in the graphics area. Program 6.10 demonstrates the use of the Apple commands by plotting the data from Table 3.1.

6.3.4 Double Density Plotting

As stated in Section 6.3.2, the Commodore 4000 and 8000 series microcomputers have only the character mode available, yet by use of 'pseudo-pixel' characters a 40 column Commodore microcomputer can produce a resolution of 80 × 50. This is on a par with the low resolution mode of an Apple. Each character position can be divided into four by certain graphics characters giving the 80 × 50 resolution (Fig. 6.9). Unfortunately a problem arises if two or more data points plot in the same character position. For example, two data points may plot as Fig. 6.9a and b in the same character location. But there are no commands that can switch on only part of a character location as on Apple II and TRS-80. To plot points Fig. 6.9a and 6.9b in the same location they must be combined to produce a new graphics character which is then plotted (Fig. 6.10). When plotting data

Fig. 6.9 Pseudo-pixel characters (Commodore).

Fig. 6.10 Combination of pseudo-pixel characters.

by this method it is necessary to know if anything has been previously plotted so that data points are not overwritten. This can be achieved in two ways. The first would be to store all the data points and before plotting check through them to find if any would overwrite. This is rather time consuming and a better method is to use the PEEK and POKE statements. Since the screen information is stored in memory we use the PEEK statement to examine any screen position to determine what, if anything, has been plotted. Having done this the appropriate graphics character can be determined and then poked into the screen memory. This technique saves machine time and storage. The graphics characters used in double density plotting, together with their POKE values are given in Table 6.4.

Table 6.4 Pseudo-Pixel Characters

Character number	Poke value	Graphics character
0	32	
1	108	
2	123	
3	98	
4	124	
5	225	
6	255	
7	254	
8	126	
9	127	
10	97	
11	252	
12	226	
13	251	
14	236	
15	160	

In order to plot in double density the following steps should be followed:

Step 1 Convert data to a value between 0 and 80 for horizontal values (x) and 0 to 50 for vertical values (y) according to the scales chosen.

Step 2 Calculate the screen address and Peek the screen memory to obtain anything that is already there.

Step 3 Find where the datum point is to be placed within the 2 × 2 grid of the character position and combine this with the information found in Step 2.

Step 4 Poke the new character on the screen at the address calculated in Step 2.

Program 6.11 demonstrates this method of double density plotting. Let us consider each step in turn.

(a) Step 1. Using the same data as the low resolution plot example, a series of statements can be written requesting the data, together with the horizontal and vertical limits. The limits are used to obtain a scaling factor to scale the data, lines 25Ø to 39Ø in Program 6.11. Entering the limits and calculating the scaling factor produces a more versatile program and therefore care must be taken when drawing the axes. In Program 6.11 the axes are positioned for the following limits:

low x = −4; high x = 76; range 80
low y = −4; high y = 46; range 50

(b) Step 2. The previous step converts the data into values between 0 and 80 for the liquid limit values and 0 and 50 for the plastic limit values. The following formula converts these values to obtain the screen address:

Screen Address (SA) = 32768 + INT (LL (I)/2) + 40*INT(25-INT(PL (I)/2))

The 25-INT(PL (I)/2) in the formula ensures that positive and negative values are plotted correctly and conventionally.

(c) Steps 3 and 4. Step 3 is probably the most difficult of all, but by using the Boolean OR statement it can be accomplished. First let us examine one character position. It is essentially divided into a 2 × 2 grid and we can think of each graphics character to be a combination of the individual quarters. If we represent a quarter segment when it is to be displayed as 1 and a zero when it is not, then each character can be represented by a 4 bit pattern. These bit patterns are shown in Table 6.5. In binary these bit patterns represent the numbers 0 to 15 which gives us the character number in Tables 6.4 and 6.5.

When a datum point is plotted we must first determine which one of the four segments in the single character position is switched on; this is achieved in lines 79Ø to 86Ø. Having determined which segment will be switched on and also its character number

Table 6.5 Pseudo-Pixel Bit Patterns

Single Character Quarters

1	2
3	4

Bit Patterns

Character no.	Graphics character	Quarters 1	2	3	4
0		0	0	0	0
1		0	0	0	1
2		0	0	1	0
3		0	0	1	1
4		0	1	0	0
5		0	1	0	1
6		0	1	1	0
7		0	1	1	1
8		1	0	0	0
9		1	0	0	1
10		1	0	1	0
11		1	0	1	1
12		1	1	0	0
13		1	1	0	1
14		1	1	1	0
15		1	1	1	1

Program 6.11

```
10 REM ************************************
20 REM    DOUBLE DENSITY PLOTTING
30 REM
40 REM  THIS PROGRAM WILL RUN ON A COMMODORE
50 REM  MICRO-COMPUTER 4000 SERIES. IT CAN
60 REM  BE MODIFIED TO ALSO RUN ON THE
70 REM  COMMODORE 8000 SERIES MACHINES
80 REM
90 REM ************************************
100 REM   CURSOR CONTROL CHARACTERS ARE
110 REM   SHOWN AS FOLLOWS :
120 REM     [CLR] ......CLEAR SCREEN
130 REM     [HOM] ......HOME CURSOR
140 REM     [CD]  ......CURSOR DOWN
150 REM     [CU]  ......CURSOR UP
160 REM     [CR]  ......CURSOR RIGHT
170 REM     [CL]  ......CURSOR LEFT
180 REM     [RVO] ......REVERSE VIDEO ON
190 REM     [RVF] ......REVERSE VIDEO OFF
200 REM   TO INDICATE MORE THAN ONE OF THE
210 REM   ABOVE A NUMBER IS PLACED WITHIN
220 REM   THE BRACKETS : I.E. [6CD]
230 REM ****** INPUT DATA *************
240 PRINT"[CLR][2CD][7CR][RVO] DOUBLE DENSITY PLOTTING"
250 DIM LL(4),PL(4)
260 PRINT"[2CD] ENTER DATA"
270 FOR I=1 TO 4
280 PRINT"[CD] POINT";I
290 INPUT" LIQUID LIMIT";LL(I)
300 INPUT" PLASTIC LIMIT";PL(I)
310 NEXT I
320 INPUT"[2CD]ENTER LOW X, HIGH X";LX,HX
330 INPUT"[CD]ENTER LOW Y, HIGH Y";LY,HY
340 REM **** OBTAIN SCALE FACTOR *****
350 PRINT"[CLR]"
360 XS=80/(HX-LX)
370 YS=50/(HY-LY)
380 OX=0-LX
390 OY=0-LY
400 FOR I=32769 TO 33649 STEP 40
410 POKE I,103
420 NEXT I
430 FOR I=33690 TO 33727
440 POKEI,99
450 NEXT I
460 PRINT"[HOM][23CD][11CR] LIQUID LIMIT"
470 PRINT"[HOM][4CD]P"
```

```
480 PRINT"L"
490 PRINT"A"
500 PRINT"S"
510 PRINT"T"
520 PRINT"I"
530 PRINT"C"
540 PRINT" "
550 PRINT"L"
560 PRINT"I"
570 PRINT"M"
580 PRINT"I"
590 PRINT"T"
600 REM **** SCALE DATA *************
610 FOR I=1 TO 4
620 PL(I)=PL(I)*YS+OY
630 LL(I)=LL(I)*XS+OX
640 NEXT I
650 DIM CH(15)
660 FOR I=0 TO 15
670 READ CH(I)
680 NEXT I
690 DATA 32,108,123,98,124,225,255,254
700 DATA 126,127,97,252,226,251,236,160
710 FOR I=1 TO 4
720 SA=32768+INT(LL(I)/2)+40*INT(25-INT(PL(I)/2))
730 PK=PEEK(SA)
740 EC=0
750 FOR J=0 TO 15
760 IF PK=CH(J) THEN EC=J:GOTO780
770 NEXT J
780 REM **** END OF ROUTINE *********
790 EX=LL(I)-2*INT(LL(I)/2)
800 EY=PL(I)-2*INT(PL(I)/2)
810 SQ=INT(EX+EY+.5)
820 ON SQ GOTO 830,840,850,860
830 CN=2 :GOTO 870
840 CN=1 :GOTO 870
850 CN=8 :GOTO 870
860 CN=4
870 REM ****END OF ON GOTO **********
880 PO= EC OR CN
890 POKE SA,CH(PO)
900 NEXT I
910 GET A$
920 IF A$="" THEN 910
930 END
```

(variable *CN* in the program) we can now PEEK the screen address where this character is to be poked to obtain the character that may already exist. The character (if any) existing at the screen address is determined in lines 65Ø to 77Ø. Here the possible plotting characters are read into a data table (lines 65Ø to 7ØØ), the screen address peeked and the character found compared with those in the table. When a match is found the existing character number (variable *EC*) is set to the appropriate value (0 to 15).

Having found the existing character number (*EC*) and the character number of the data point (*CN*) we use the OR statement to give us the correct character number to be plotted. Commodore BASIC supports the Boolean operators: AND, OR and NOT. They are used mainly with IF statements and their operation is best explained by the following examples.

(a) IF *A* = *C* AND *B* = *C* THEN GOTO 3ØØ
A jump to line 3ØØ will occur only if both *A* and *B* equal *C*.

(b) IF *A* = *C* OR *B* = *C* THEN GOTO 3ØØ
The jump will now occur if either of *A* or *B* is equal to *C*.

(c) IF NOT *A* THEN GOTO 2ØØ
The branch to line 2ØØ will occur if *A* is equal to zero.

These Boolean operators can also be used in an assignment statement, for example,

A = 43 AND 137

In such a form the computer operates on these numbers bit by bit and it is in this way that the OR statement is utilized.

Let us assume that the existing character number (*EC*) is 1 and the character number of the data to be plotted (*CN*) is 4. The combination of these characters will produce the new character to be plotted (Fig. 6.10). It may seem that simple addition of the character numbers will give us the character number of the character to be plotted but this is not always the case. The only way to obtain the correct character number is to apply the operator OR to the character numbers. The OR statement takes the bit patterns for the values assigned to *EC* and *CN* and combines them to produce a new bit pattern. The OR operation results in a 1 if either bit is 1 in the bit patterns OR'd together. Thus, if 1 and 4 are OR'd together the resultant bit pattern is for 5, for example

ONE	OR	FOUR	=	FIVE
0001		0100		0101

Some additional examples are given in Table 6.6 for the reader to examine and convince himself that the correct character number is always produced. This process is achieved on line 88Ø of Program 6.11, and all that is left is for the point to be plotted by the POKE statement on line 89Ø.

The graph can then be embellished with axes and labels; lines 4ØØ to 59Ø. The output (Fig. 6.11) of Program 6.11 is a line printer copy of the screen display.

6.3.5 High Resolution

Certain microcomputers have a high resolution capability, for example Apple II, Apple III, Atari (400 and 800), the BBC and Acorn Atom microcomputers. High resolution graphics enable accurate graphs, realistic

Table 6.6 Generation of Updated Character (additional examples)

	Existing Character	OR'd	New Character	=	Updated Character
(a)				=	
	0011		0100		0111
(b)				=	
	1011		0010		1011

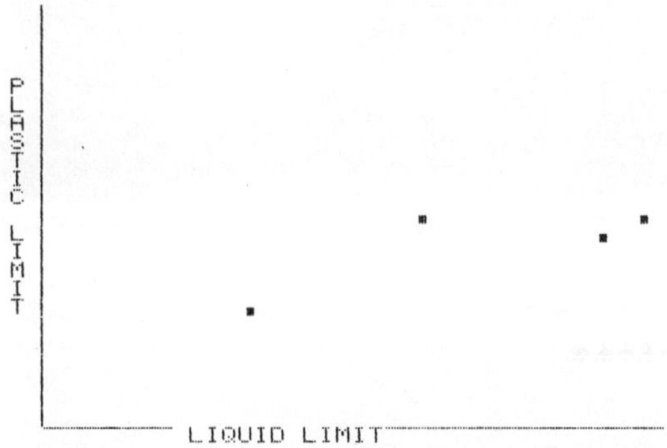

Fig. 6.11 Screen output from Program 6.11.

diagrams and three-dimensional drawings to be displayed and used, particularly in design-type programs.

The Apple II has a resolution of 280 horizontal positions by 192 vertical positions with the point 0,0 in the top left-hand corner. Utilizing the high resolution capabilities does have its difficulties and this topic will only be briefly covered.

On an Apple, high resolution is activated by issuing the command:

HGR

either directly from the keyboard or from a program. The command sets up an area of memory for high resolution graphics and clears it. The HGR instruction will only allow 160 of the 192 vertical high resolution points to be displayed; the bottom portion of the screen being used as a text window. In order to display the full screen the following POKE instruction must be issued after the HGR.

POKE — 16302,0.

This eliminates the text window. The basic command in this mode is the HPLOT instruction and can be used in one of the following three ways:

(a) HPLOT *X*,*Y* This plots a single point on the screen.
(b) HPLOT *X*1, *Y*1 to *X*2, *Y*2 This draws a line from point *X*1, *Y*1 to *X*2, *Y*2; for example HPLOT 0,0 TO 279, 192 plots a line from the upper left-hand corner to the lower right-hand corner.
(c) HPLOT *X*1, *Y*1 to *X*2, *Y*2 to *X*3, *Y*3 This joins a series of points together; *X*1, *Y*1 to *X*2, *Y*2 to *X*3, *Y*3.

There are additional commands which can be used in the high resolution mode and the reader should consult the Apple microcomputer manuals.

High resolution graphics is an extensive subject in its own right. Three-dimensional diagrams with shading and animation are possible on the Apple II and this is obviously beyond the scope of this text. A simple demonstration program for Apple II is shown in Program 6.12. The program accepts data for two shapes: a hexagon and a square; it then calculates a series of new intermediate points to convert the hexagon to a square and then joins up the main points. It is a process that forms a series of multiple images.

6.4 Hard Copy

Output to the screen, no matter what the resolution, is only temporary. It is, therefore, necessary to be able to have a permanent record (hard copy) of screen output. Normal text output can be recorded on printers which can have a quality of print varying from standard dot matrix head output to letter (electric typewriter) quality. The dot matrix head printers produce characters by firing small needles on a grid (similar to screen character formation). The needles impinge on the printer ribbon and so display characters on the paper. On some printers this type of printing head is also capable of reproducing low or high resolution graphics drawn on a VDU (Fig. 6.9). A sample of graphical output from a Commodore 4022 printer can be seen in Fig. 11.5. The standard of copy, although adequate for many purposes, can be improved if an *X*–*Y* plotter is used.

An *X*–*Y* plotter consists of a pen which can be moved over a piece of paper under the control of a program. The pen can be instructed

Program 6.12

```
10 REM *****************************
20 REM    APPLESOFT BASIC
30 REM   HIGH RESOLUTION
40 REM *****************************
50 DIM X(2,10),Y(2,10)
60 FOR I=1 TO  2
70 FOR NA=1 TO 7
80 READ X(I,NA),Y(I,NA)
90 NEXT NA
100 NEXT I
110 REM **CLEAR SCREEN-SET HIGH RES.**
120 HOME
130 HGR
140 POKE -16302,0
150 HCOLR = 7
160 FOR I=1 TO 7-1
170 HPLOT X(1,I),Y(1,I) TO X(1,I+1),Y(1,I+1)
180 NEXT I
190 FOR I=1 TO 10
200 FOR P=0 TO I-1
210 SP=SP+1
220 NEXT P
230 FOR J=1 TO NA-2
240 X1=X(1,J)+(SP*(X(2,J)-X(1,J)))
250 Y1=Y(1,J)+(SP*(Y(2,J)-Y(1,J)))
260 X2=X(1,J+1)+(SP*(X(2,J+1)-X(1,J+1)))
270 Y2=Y(1,J+1)+(SP*(X(2,J+1)-X(1,J+1)))
280 HPLOT X1,Y1 TO X2,Y2
290 NEXT J
300 SP=0
310 NEXT I
320 FOR I=1 TO NA
330 HPLOT X(1,I),Y(1,I) TO X(2,I),Y(2,I)
340 NEXT I
350 REM *****  DATA *****************
360 REM HEXAGON DATA
370 DATA 50,0,25,25,25,75,50,10,75,75,75,25,50,0
380 REM SQUARE DATA
390 DATA 225,125,200,125,200,175,225,175,250,175,250,125,225,125
400 END
```

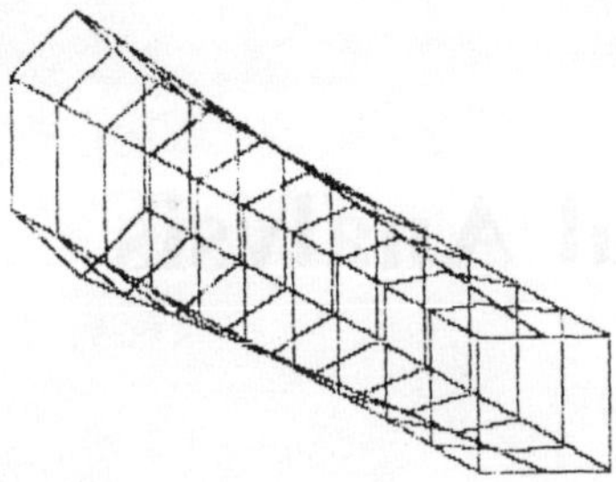

Fig. 6.12 High-resolution screen copy of output of Program 6.12.

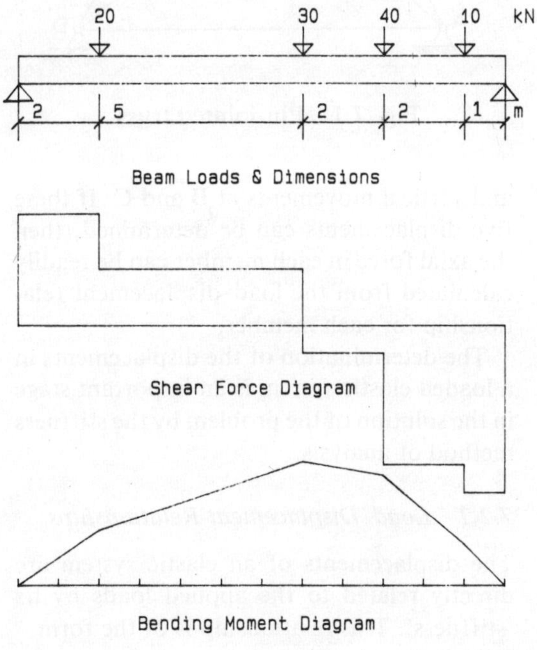

Fig. 6.13

to be lifted from, or placed on the paper and moved in either the *X* or *Y* direction. Combining movements in the *X* and *Y* directions enables curves, sloping lines and circles to be drawn. Realistic curves will only be drawn if the steps in the *X* and *Y* directions are small thus the resolution of *X–Y* plotters should be in the region of 0.1 mm.

Many plotters have their own microprocessors which accept their own special commands, these are known as 'intelligent' plotters. Single commands allow complex shapes to be drawn and even pens to be changed automatically.

An example of output from an *X–Y* plotter under microcomputer control is shown in Fig. 6.13. As can be seen the standard of output is high. Programs can therefore be written to produce not only graphs but such things as detailing of reinforcement for reinforced concrete structural elements or surveying traverses, etc.

Further Reading

Korites, B. J. *Graphic Software for Microcomputers*. Kern Publications, 1981.

Newman, W. M. and Sproull, R. F. *Principles of Interactive Computer Graphics*, 5th edn. McGraw-Hill, 1979.

7

Applications in Structural Analysis

7.1 Introduction

As the influence of the computer has become more widespread, it has become clear that the most useful general approach to structural analysis in the context of the computer is by the 'stiffness' method. Most analysis done on computers is based on the stiffness method, so in this chapter we shall concentrate on this method. We shall lead up to two general programs, one to analyse pin-jointed plane trusses, and another to analyse rigidly jointed plane frames.

First we look at the fundamentals of the stiffness method.

7.2 The Stiffness Method for Plane Truss Analysis

The following is an outline development only. For a full explanation, the reader is advised to consult a modern text in structural analysis[1, 2, 3]. In the stiffness method, attention is focused on the displacements which develop at the joints or 'nodes' of the structure.

In the pin-jointed truss shown in Fig. 7.1, there are four joints A, B, C, and D connecting the six members. The support at A is pinned, thus providing horizontal and vertical components of reaction and preventing any linear movements at this joint. Joint D is on a roller support and this allows horizontal movement of this joint to take place. There are thus five independent joint displacements: horizontal movements at B, C, and D, and vertical movements at B and C. If these five displacements can be determined, then the axial force in each member can be readily calculated from the load–displacement relationship for each member.

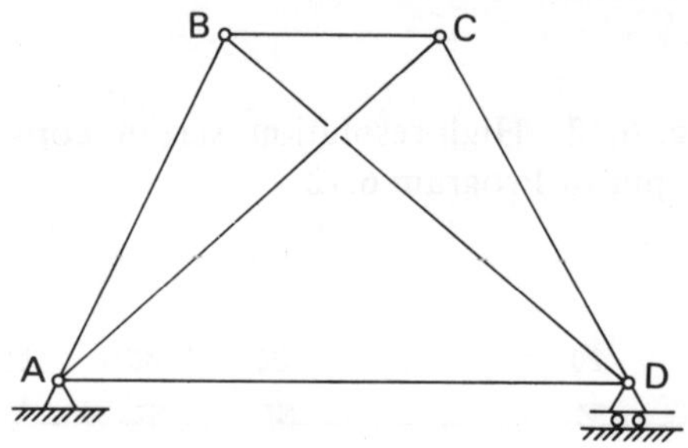

Fig. 7.1 Pin-jointed truss.

The determination of the displacements in a loaded elastic system is an important stage in the solution of the problem by the stiffness method of analysis.

7.2.1 *Load–Displacement Relationships*

The displacements of an elastic system are directly related to the applied loads by its 'stiffness'. This relationship is of the form

$$P = K\Delta \tag{7.1}$$

where P = load, Δ = displacement and K = stiffness. In the simplest case, say that of a spring which deforms by Δ under the load P, (P/Δ) is the spring stiffness. This is a single-valued number which represents the force required to produce unit displacement. In most structures, there are normally several displacements and forces involved. In such

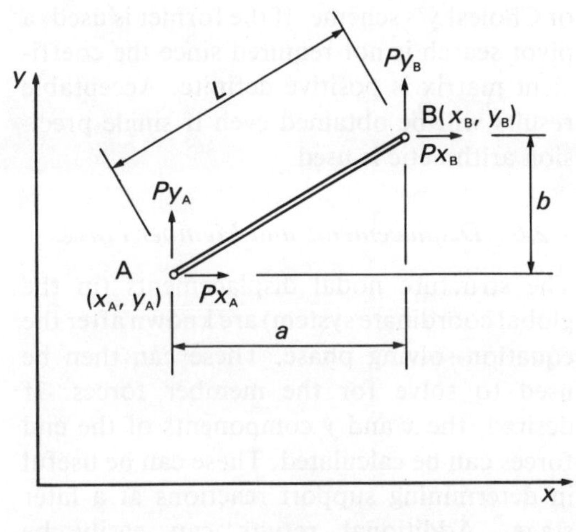

Fig. 7.2 Truss member.

cases, the load–displacement relationship is still expressed in the form of Equation 7.1 but each term will have to be interpreted as a matrix.

7.2.2 *Stiffness Matrix of a Truss Member*

A typical truss member A–B (Fig. 7.2) is assumed to be pin-connected to the rest of the structure and thus only axial force may be developed within the member. Each end of the member may undergo linear displacements and it is usual to consider the components of the displacements in two perpendicular directions. Normally, these are taken to be in the *x* and *y* directions. For each displacement there corresponds a linear force and the load–displacement relationship takes the form shown in Equation 7.2.

$$\begin{bmatrix} Px_A \\ Py_A \\ Px_B \\ Py_B \end{bmatrix} = \begin{bmatrix} K_{11} & K_{12} & K_{13} & K_{14} \\ & K_{22} & K_{23} & K_{24} \\ \text{(SYM.)} & & K_{33} & K_{34} \\ & & & K_{44} \end{bmatrix} \begin{bmatrix} \Delta x_A \\ \Delta y_A \\ \Delta x_B \\ \Delta y_B \end{bmatrix} \quad (7.2)$$

By allowing each of the displacements to take unit value in turn, while constraining all other displacements to zero, the elements of the stiffness matrix can be determined.

Thus, with the notation shown on Fig. 7.2, matrix *K* (Equation 7.2) for a truss member can be shown to be:

$$K = \frac{AE}{L^3} \begin{bmatrix} a^2 & ab & -a^2 & -ab \\ & b^2 & -ab & -b^2 \\ \text{(SYM.)} & & a^2 & ab \\ & & & b^2 \end{bmatrix} \quad (7.3)$$

The matrix shown in Equation 7.3 relates the member forces and displacements with reference to the *x–y* axes.

7.2.3 *Stiffness Matrix of a Complete Structure*

For an elastic structure, the load–deflection relationship can also be represented by Equation 7.1. The matrix *K* is the 'overall stiffness' matrix; the elements in each column are by definition the nodal forces which correspond to unit value of each displacement, with all other displacements held to zero.

For the computer analysis of a given structure, it is highly desirable that each step in the solution be capable of being programmed efficiently. The assembly of the stiffness matrix *K* is of particular importance and is of considerable programming interest. This may be done in one of many ways. Some methods require the calculation of triple matrix products, but such methods require not only relatively long computing time, but also considerable storage capacity within the computer. For small desk-top machines, this method is not very suitable.

A simple way of assembly is by direct addition of stiffnesses and this will be described in the next section.

7.2.4 *Assembly of Stiffness Matrix*

The stiffness matrix of a typical truss member is given by Equation 7.3. Now, each member will contribute to the overall stiffness matrix of the whole structure and could be thought of as the 'building block' of the matrix. The contribution of any member is effected by calculating the member stiffness matrix and transferring the elements into the appropriate addresses of the structure stiffness matrix.

The correct destination addresses are easily determined by following a cyclic permutation of the row and column numbers (1–4 for truss members, 1–6 for rigid frame members) which have a direct correspondence with the structure displacement numbers.

At any support joint where one or more displacements are constrained to zero, a 0 value is assigned as the 'displacement number'. This would require that the structure stiffness matrix be augmented by column number 0 and row number 0 so that these may be used as a 'dump' for all stiffnesses that are not required. In the PET BASIC system, arrays are automatically given a starting address of 0 in the dimension declaration and therefore this requirement is catered for. This makes the programming considerably simpler. If desired, however, the program can be modified so that:

(a) 'dump' locations become unnecessary; and
(b) all array addresses start at 0

thus gaining a small amount of storage space. However, the gain in additional storage would be small in most cases and the additional programming effort is not justified. The reader may nevertheless wish to make this programming modification as an exercise.

It may be pointed out here that a number of other modifications can be made to maximize the utilization of available storage space in the random access memory. Most obvious are those that take advantage of the symmetry and banded form of the structure stiffness matrix. Details of these are outside the scope of this present work.

7.2.5 Solution of Equilibrium Equations

After the applied load vector has been read in, the solution of the equations takes place using a suitable method such as that of Gauss or Cholesky's scheme. If the former is used, a pivot search is not required since the coefficient matrix is positive definite. Acceptable results will be obtained even if single precision arithmetic is used.

7.2.6 Displacements and Member Forces

The structure nodal displacements (in the global coordinate system) are known after the equation-solving phase. These can then be used to solve for the member forces. If desired, the x and y components of the end forces can be calculated. These can be useful in determining support reactions at a later stage. Additional results can easily be obtained by suitable alteration to the program (e.g. axial stress, member deformation).

7.2.7 The Plane Truss Analysis Program

Program 7.1 will analyse plane trusses in accordance with the principles just outlined. A printer is assumed to be available to receive the output, but the user can modify the program to suit his particular computing environment.

The structure of the program is such that details of the truss to be analysed are given in 'data' statements typed in at appropriate positions. The beginning of the program is perhaps the most convenient place since the line numbers can begin at 1. This approach to the input of data has the advantage that a structure can be modified and reanalysed without having to re-enter all the data.

The modulus of elasticity is assumed to be constant for each individual structure. This is the case with the great majority of problems, but on the rare occasions where the members have different values for this parameter, it would not be difficult to modify the program to suit.

7.2.8 Data Preparation

An x–y coordinate system is used for ease of data preparation. It is recommended that the origin be located such that all joint coordi-

Program 7.1 Plane truss analysis — program and output

```
5 REM DATA
10 DATA 1,11,10
11 DATA 0,1.732,1,1.732,10,0,0,1,2
12 DATA 1,1.732,3,1.732,10,1,2,3,4
13 DATA 0,1.732,.5,.866,10,0,0,5,6
14 DATA .5,.866,1,1.732,10,5,6,1,2
15 DATA 1,1.732,1.5,.866,10,1,2,7,8
16 DATA 1.5,.866,3,1.732,10,7,8,3,4
17 DATA .5,.866,1.5,.866,10,5,6,7,8
18 DATA 0,0,.5,.866,10,0,0,5,6
19 DATA .5,.866,1,0,10,5,6,9,10
20 DATA 1,0,1.5,.866,10,9,10,7,8
21 DATA 0,0,1,0,10,0,0,9,10
22 DATA 0,-1,0,-1,0,0,0,0,0,0
100 OPEN 1,4
110 OPEN 2,4,1
120 OPEN 3,4,2
130 FOR I=1 TO 80
140 D$=D$+"*"
150 NEXT
160 GOSUB 5000
170 PRINT#1,"TRUSS ANALYSIS"
180 GOSUB 5000
185 REM READ DATA
190 READ E,NM,N
200 PRINT#1,"MODULUS OF ELASTICITY=";E
210 PRINT#1,"NUMBER OF MEMBERS=";NM
220 PRINT#1,"NUMBER OF UNKNOWN DISPLACEMENTS=";N
230 GOSUB 5000
240 DIM P(NM,9),A(N,N),SB(N),B(N),X(N),Q(NM,4),M(4,4)
250 FOR I=1 TO NM
260 READ P(I,1),P(I,2),P(I,3),P(I,4),P(I,5),Q(I,1),Q(I,2),Q(I,3),Q(I,4)
270 P(I,7)=P(I,3)-P(I,1)
280 P(I,8)=P(I,4)-P(I,2)
290 P(I,6)=SQR(P(I,7)↑2+P(I,8)↑2)
300 NEXT I
310 FOR I=1 TO N
320 READ B(I)
330 SB(I)=B(I)
340 NEXT I
345 REM ASSEMBLE STRUCTURE STIFFNESS MATRIX
350 FOR R=1 TO NM
360 GOSUB 10000
370 FOR I=1 TO 4
380 FOR J=1 TO 4
390 A(Q(R,I),Q(R,J))=A(Q(R,I),Q(R,J))+M(I,J)
400 NEXT J,I,R
```

```
405 REM SOLVE EQUATIONS
410 GOSUB 20000
420 F$="99   S999.999   S999.9999999"
430 PRINT#3,F$
440 PRINT#1,"NO.   LOAD   DISPLACEMENT"
450 GOSUB 5000
455 REM PRINT DISPLACEMENTS
460 FOR R=1 TO N
470 PRINT#1
480 PRINT#2,R,SB(R),X(R)
490 NEXT R
500 GOSUB 5000
505 REM PRINT MEMBER FORCES
510 PRINT#1,"NO.    XA      YA      XB      YB      AREA    LENGTH    FORCE"
520 GOSUB 5000
530 F$="99  S99.999  S99.999  S99.999  S99.999  99999  999.999   S999.9999"
540 PRINT#3,F$
550 FOR R=1 TO NM
560 P(R,9)=(X(Q(R,3))-X(Q(R,1)))*P(R,7)+(X(Q(R,4))-X(Q(R,2)))*P(R,8)
570 P(R,9)=P(R,9)*(P(R,5)*E/P(R,6)↑2)
580 PRINT#1
590 PRINT#2,R,P(R,1),P(R,2),P(R,3),P(R,4),P(R,5),P(R,6),P(R,9)
600 NEXT R
605 GOSUB 5000
610 GOTO 30000
5000 PRINT#1
5010 PRINT#1
5020 PRINT#1,D$
5030 PRINT#1
5040 PRINT#1
5050 RETURN
9000 REM SUBROUTINE TO CALCULATE MEMBER STIFFNESS MATRIX
10000 A2=P(R,7)↑2
10010 B2=P(R,8)↑2
10020 AB=(P(R,7)*P(R,8))
10030 M(1,1)=A2
10040 M(3,3)=A2
10050 M(1,3)=-A2
10060 M(3,1)=-A2
10070 M(2,2)=B2
10080 M(4,4)=B2
10090 M(2,4)=-B2
10100 M(4,2)=-B2
10110 M(1,2)=AB
10120 M(2,1)=AB
10130 M(3,4)=AB
10140 M(4,3)=AB
10150 M(1,4)=-AB
10160 M(2,3)=-AB
```

```
10170 M(3,2)=-AB
10180 M(4,1)=-AB
10190 FOR I=1 TO 4
10200 FOR J=1 TO 4
10210 M(I,J)=M(I,J)*P(R,5)*E/P(R,6)↑3
10220 NEXT J,I
10230 RETURN
19000 REM SUBROUTINE TO SOLVE EQUATIONS
20000 FOR I=1 TO N-1
20010 FOR J=I+1 TO N
20020 D=A(J,I)/A(I,I)
20030 FOR K=1 TO N
20040 A(J,K)=A(J,K)-D*A(I,K)
20050 NEXT K
20060 B(J)=B(J)-D*B(I)
20070 NEXT J,I
20080 X(N)=B(N)/A(N,N)
20090 FOR I=1 TO N-1
20100 J=N-I
20110 D=0
20120 FOR K=J+1 TO N
20130 D=D+A(J,K)*X(K)
20140 NEXT K
20150 X(J)=B(J)/A(J,J)-D/A(J,J)
20160 NEXT I
20170 RETURN
30000 END
```

```
TRUSS ANALYSIS
```

```
MODULUS OF ELASTICITY= 1
NUMBER OF MEMBERS= 11
NUMBER OF UNKNOWN DISPLACEMENTS= 10
```

NO.	LOAD	DISPLACEMENT
1	+ .000	+ .1924552
2	- 1.000	- .3777802
3	+ .000	+ .5388755
4	- 1.000	- 2.4262446
5	+ .000	- .0769751
6	+ .000	- .1333323
7	+ .000	- .1539569
8	+ .000	- .5333426
9	+ .000	- .1539650
10	+ .000	- .3555608

NO.	XA	YA	XB	YB	AREA	LENGTH	FORCE
1	+ .000	+ 1.732	+ 1.000	+ 1.732	10	1.000	+ 1.9245
2	+ 1.000	+ 1.732	+ 3.000	+ 1.732	10	2.000	+ 1.7321
3	+ .000	+ 1.732	+ .500	+ .866	10	.999	+ .7698
4	+ .500	+ .866	+ 1.000	+ 1.732	10	.999	- .7698
5	+ 1.000	+ 1.732	+ 1.500	+ .866	10	.999	- .3849
6	+ 1.500	+ .866	+ 3.000	+ 1.732	10	1.732	- 2.0000
7	+ .500	+ .866	+ 1.500	+ .866	10	1.000	- .7698
8	+ .000	+ .000	+ .500	+ .866	10	.999	- 1.5396
9	+ .500	+ .866	+ 1.000	+ .000	10	.999	+ 1.5396
10	+ 1.000	+ .000	+ 1.500	+ .866	10	.999	- 1.5396
11	+ .000	+ .000	+ 1.000	+ .000	10	1.000	- 1.5396

nates are positive. However, this is not essential.

For each member, the 'member direction' will have to be specified, i.e. the direction from joint A to joint B. An arrowhead drawn along the member length will suffice for this purpose.

Members are numbered consecutively from 1 to NM (total number of members). Joint displacements will also have to be identified and numbered consecutively with the x and y displacements numbered in that order at each joint. At supports, suppressed displacements are to be designated by 0.

Having carried out these preliminary steps, the data are now typed in. Starting at the first line number, we type in:

DATA E, NM, N

For the structure being analysed. After the next line number, we type in:

DATA x_A, y_A, x_B, y_B, A, d_1, d_2, d_3, d_4

For member number 1, where E = modulus of elasticity, NM = number of members, N = number of unknown displacements, x_A, y_A, x_B, y_B are the coordinates of end A and end B respectively, A = cross-sectional area of the member, and d_1, d_2, d_3 and d_4 are the displacement numbers at end A and end B.

Data for the remaining members follow consecutively in the same pattern. The last line of the data statements takes the form:

DATA P_1, P_2, P_3 . . . P_N

where P_1, P_2, P_3, . . ., P_N are the nodal loads.

Example 7.1 Plane Truss Analysis

Figure 7.3a shows a pin-jointed truss to be analysed for the member forces under the action of the joint loads shown. The modulus of elasticity is taken equal to 1, and all members have equal cross-sectional areas. Although we note that this truss is statically indeterminate, this fact is irrelevant in applying the stiffness method.

Figs. 7.3b and 7.3c show the numbering of

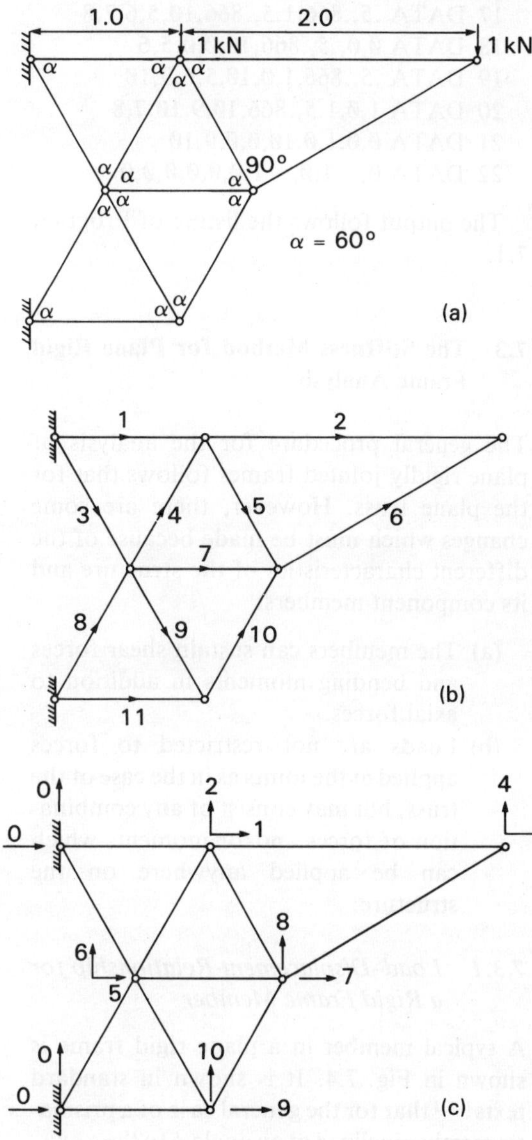

Fig. 7.3

the members, member directions, and the numbering of the displacements.

The data are typed in as follows:

```
1Ø DATA 1,11,1Ø
11 DATA Ø,1.732,1,1.732,1Ø,Ø,Ø,1,2
12 DATA 1,1.732,3,1.732,1Ø,1,2,3,4
13 DATA Ø,1.732,.5,.866,1Ø,Ø,Ø,5,6
14 DATA .5,.866,1,1.732,1Ø,5,6,1,2
15 DATA 1,1.732,1.5,.866,1Ø,1,2,7,8
16 DATA 1.5,.866,3,1.732,1Ø,7,8,3,4
```

```
17 DATA .5,.866,1.5,.866,10,5,6,7,8
18 DATA 0,0,.5,.866,10,0,0,5,6
19 DATA .5,.866,1,0,10,5,6,9,10
20 DATA 1,0,1.5,.866,10,9,10,7,8
21 DATA 0,0,1,0,10,0,0,9,10
22 DATA 0, - 1,0, - 1,0,0,0,0,0,0
```

The output follows the listing of Program 7.1.

7.3 The Stiffness Method for Plane Rigid Frame Analysis

The general procedure for the analysis of plane rigidly jointed frames follows that for the plane truss. However, there are some changes which must be made because of the different characteristics of the structure and its component members:

(a) The members can sustain shear forces and bending moments in addition to axial forces.

(b) Loads are not restricted to forces applied at the joints as in the case of the truss, but may consist of any combination of forces and/or moments which can be applied anywhere on the structure.

7.3.1 Load–Displacement Relationship for a Rigid Frame Member

A typical member in a plane rigid frame is shown in Fig. 7.4. It is shown in standard texts[1,2,3] that for the general case of a prismatic member inclined at an angle ϕ to the x axis, the load–displacement relationship is given by Equation 7.4.

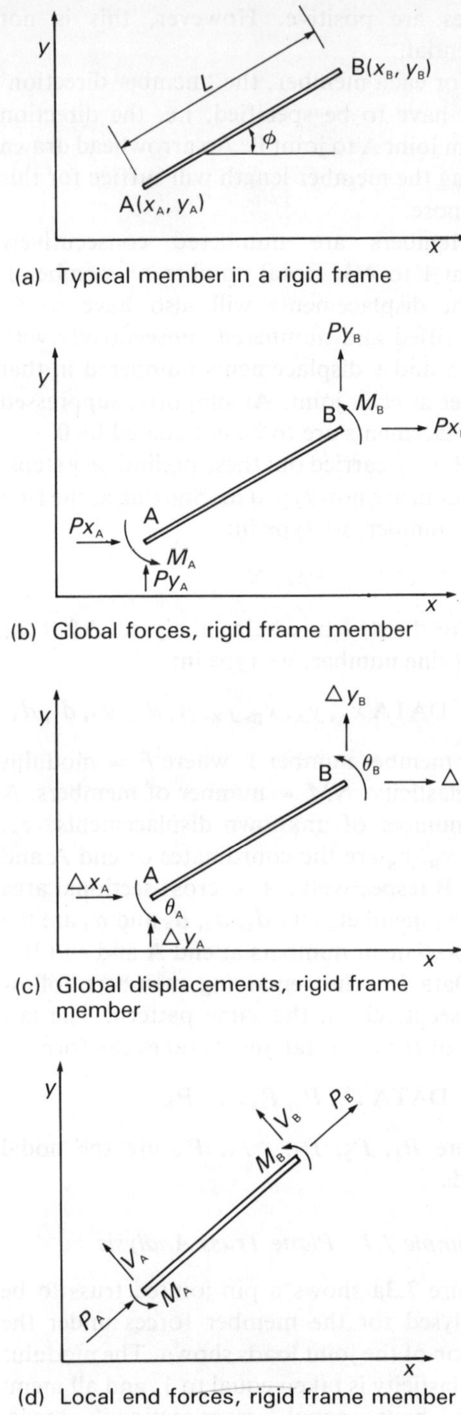

Fig. 7.4

$$
\begin{Bmatrix} P_{x_A} \\ P_{y_A} \\ M_A \\ P_{x_B} \\ P_{y_B} \\ M_B \end{Bmatrix}
=
\begin{bmatrix}
K_1 & K_2 & -K_4 & -K_1 & -K_2 & -K_4 \\
 & K_3 & K_5 & -K_2 & -K_3 & K_5 \\
 & & 2K_6 & K_4 & -K_5 & K_6 \\
 & & & K_1 & K_2 & K_4 \\
 & \text{(SYM.)} & & & K_3 & -K_5 \\
 & & & & & 2K_6
\end{bmatrix}
\begin{Bmatrix} \Delta_{x_A} \\ \Delta_{y_A} \\ \Theta_A \\ \Delta_{x_B} \\ \Delta_{y_B} \\ \Theta_B \end{Bmatrix}
\qquad (7.4)
$$

i	K_i
1	$AE\cos^2\phi/L + 12EI\sin^2\phi/L^3$
2	$(AE/L - 12EI/L^3)\sin\phi\cos\phi$
3	$AE\sin^2\phi/L + 12EI\cos^2\phi/L^3$
4	$6EI\sin\phi/L^2$
5	$6EI\cos\phi/L^2$
6	$2EI/L$

It is seen that each end of the member can have rotational as well as linear movements. The member stiffness matrix is therefore of size 6 × 6.

The sign conventions adopted for displacements are that linear movements upwards, to the right, and counterclockwise rotations are all given positive signs. The applied forces and moments also follow the same sign conventions.

7.3.2 *Treatment of Applied Loads*

As mentioned before, loads may or may not be applied at a joint. Where loads are applied at positions intermediate between the 'natural' joints of the structure, there can be two choices for the course of action:

1. The point of application of the load is designated as a joint and the displacements numbered accordingly. This may be conveniently used in cases where concentrated loads or moments are applied, and leads to a straightforward solution, but suffers the disadvantage of increasing the size of the stiffness matrix. This might be a crucial factor in certain problems, even more so when the storage capacity of the computer is small.
2. Using equivalent nodal loads, i.e. fixed end moments and shears with reversed directions, for the complementary solution, and adding the particular solution at a later stage.

The analyst has complete freedom of choice in deciding which method to adopt.

7.3.3 *Member Force Transformations*

The load–displacement system of equations for the complete structure expresses the relationships between these quantities when referred to the global axes. The solution of the equations yields the displacements from which the member forces can be calculated. These member forces are therefore with respect to the global axes and are useful for checking the overall equilibrium and to calculate the support reactions.

In so far as the members themselves are concerned, the quantities which are of immediate interest to the designer are the forces and moments in local coordinates, i.e. the axial force, shear force and bending moments. Moments are of course identical in both systems of coordinates, but the forces need to be transformed. The local forces can be obtained from the global member forces by using the relationship, Equation 7.5.

$$\begin{bmatrix} P_A \\ V_A \\ M_A \\ P_B \\ V_B \\ M_B \end{bmatrix} = \begin{bmatrix} C & S & 0 & 0 & 0 & 0 \\ -S & C & 0 & 0 & 0 & 0 \\ 0 & 0 & 1 & 0 & 0 & 0 \\ 0 & 0 & 0 & C & S & 0 \\ 0 & 0 & 0 & -S & C & 0 \\ 0 & 0 & 0 & 0 & 0 & 1 \end{bmatrix} \begin{bmatrix} P_{xA} \\ P_{yA} \\ M_A \\ P_{xB} \\ P_{yB} \\ M_B \end{bmatrix} \qquad (7.5)$$

where $C = \cos\phi$ and $S = \sin\phi$.

7.3.4 Plane Frame Program

Program 7.2 is provided for the analysis of rigid frames. This has been expressly designed to be simple, short, and easy to use. The user may wish to modify it to give it more sophistication, but the overall approach should remain the same.

As with the plane truss analysis program, the preliminary work of establishing a coordinate system, member directions, etc. has to be done first before data preparation can take place.

This program can also be used to analyse continuous beams, propped cantilevers, and similar beam-type configurations.

As a matter of interest, plane trusses can also be analysed correctly by this program if sufficiently small values for the second moment of area are given to represent members that cannot carry bending moments. This, however, is not advisable as it is wasteful of storage and computing time.

7.3.5 Data Preparation

This has a similar pattern to that for the plane truss analysis. Data should be given in the following order:

E, NM, N

$X_A, Y_A, X_B, Y_B, A, I, d_1, d_2, d_3, d_4, d_5, d_6$

(for each member)

$P_1, P_2, P_3, \ldots, P_N$

where I is the second moment of area of a member and the other symbols have the same

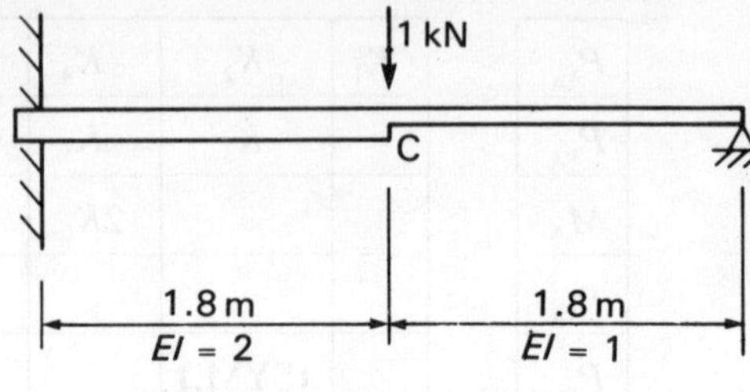

Fig. 7.5 Propped cantilever.

meaning as in the plane truss analysis program.

Two examples will be solved to illustrate the use of the program; a propped beam with non-uniform section and a rigidly jointed frame with inclined members.

Example 7.2 Propped Cantilever

Fig. 7.5 shows a propped cantilever and relevant data as to its geometry and loading condition, point C is treated as a 'joint' and the data are:

```
1 DATA 1,2,4
2 DATA Ø,Ø,1.8,Ø,1,2,Ø,Ø,Ø,1,2,3
3 DATA 1.8,Ø,3.6,Ø,1,1,1,2,3,Ø,Ø,4
4 DATA Ø, – 1,Ø,Ø
```

The results produced are shown after the listing of the program.

Example 7.3 A Plane Rigid Frame

Fig. 7.6 shows a plane rigid frame subjected to concentrated loads. All dimensions are in metres and the member properties are as shown in Table 7.1.

Table 7.1

Member	Area (m^2)	Second moment of area (m^4)
AB	3.23×10^{-3}	3.7×10^{-5}
CD	3.23×10^{-3}	5.2×10^{-5}
AC,CE	3.23×10^{-3}	2.4×10^{-5}
BD,DF	3.23×10^{-3}	2.348×10^{-5}

From these, the data for the program are prepared, and shown together with the output following the listing of the program.

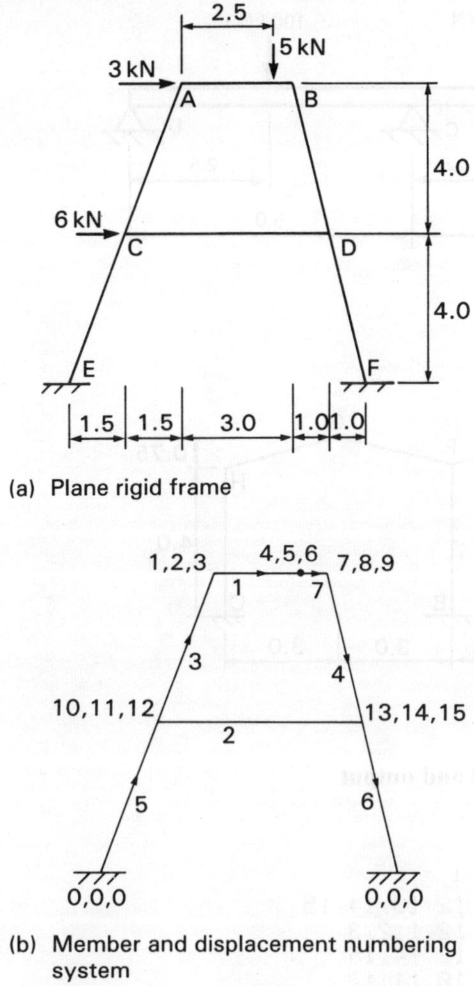

(a) Plane rigid frame

(b) Member and displacement numbering system

Fig. 7.6

Exercises

7.1 Alter Program 7.1 so that the data required are given in terms of the lengths and angles of inclination of each member. Run the program for the truss shown in Fig. 7.3a.

7.2 One half of a roof truss is shown in Fig. 7.7. A vertical downward load of 1.0 kN is applied at joints H and J. The truss and its load system are symmetrical about the centre line. Using Program 7.1, determine the axial forces in all members of the truss.

7.3 Draw the bending moment diagram for the continuous beam ABCD shown in Fig. 7.8. Use Program 7.2 to analyse the beam.

7.4 For the frame shown in Fig. 7.9, calculate all components of reactions at A,B, anc C due to the horizontal load applied at D. The members have the properties shown in Table 7.2.

Table 7.2

Member	Area × 10^3 (m^2)	Second moment of area × 10^5 (m^4)
AD,CH	3.79	2.88
BF	4.00	4.42
DE,EF, FG,GH	3.22	2.34

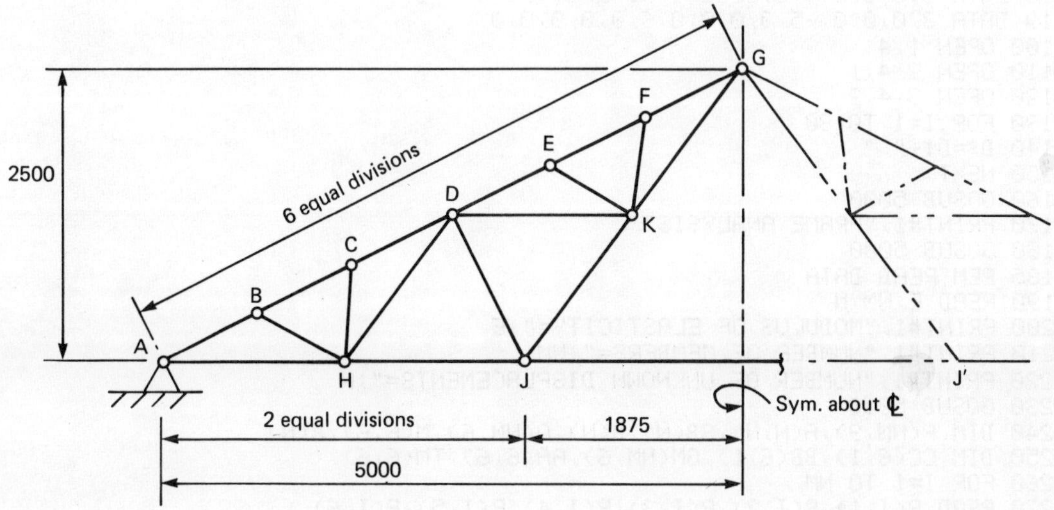

Fig. 7.7

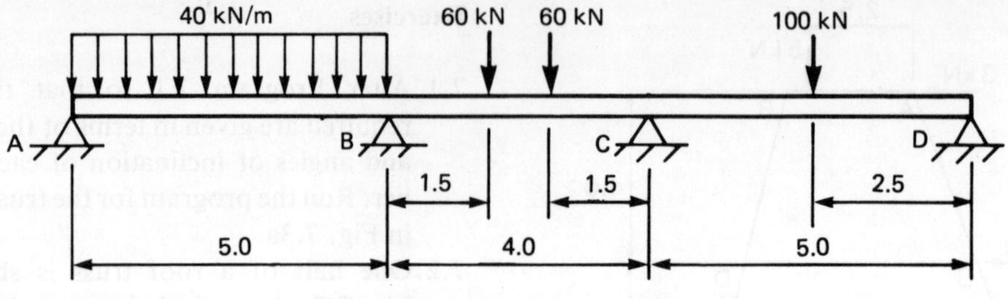

Fig. 7.8

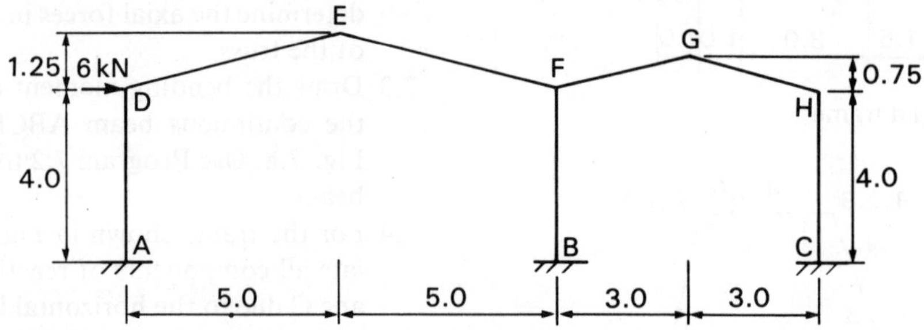

Fig. 7.9

Program 7.2 Plane rigid frame analysis — program and output

```
5 REM DATA
11 DATA 210E6,7,15
12 DATA 3,8,5.5,8,3.223E-3,3.7E-5,1,2,3,4,5,6
13 DATA 1.5,4,7,4,3.223E-3,5.2E-5,10,11,12,13,14,15
14 DATA 1.5,4,3,8,3.223E-3,2.4E-5,10,11,12,1,2,3
15 DATA 6,8,7,4,3.223E-3,2.348E-5,7,8,9,13,14,15
16 DATA 0,0,1.5,4,3.223E-3,2.4E-5,0,0,0,10,11,12
17 DATA 7,4,8,0,3.223E-3,2.348E-5,13,14,15,0,0,0
18 DATA 5.5,8,6,8,3.223E-3,3.7E-5,4,5,6,7,8,9
19 DATA 3,0,0,0,-5,0,0,0,0,6,0,0,0,0,0
100 OPEN 1,4
110 OPEN 2,4,1
120 OPEN 3,4,2
130 FOR I=1 TO 80
140 D$=D$+"*"
150 NEXT
160 GOSUB 5000
170 PRINT#1,"FRAME ANALYSIS"
180 GOSUB 5000
185 REM READ DATA
190 READ E,NM,N
200 PRINT#1,"MODULUS OF ELASTICITY=";E
210 PRINT#1,"NUMBER OF MEMBERS=";NM
220 PRINT#1,"NUMBER OF UNKNOWN DISPLACEMENTS=";N
230 GOSUB 5000
240 DIM P(NM,9),A(N,N),SB(N),B(N),Q(NM,6),M(6,6),X(N)
250 DIM CC(6,1),BB(6,1),GM(NM,6),AA(6,6),TM(6,6)
260 FOR I=1 TO NM
270 READ P(I,1),P(I,2),P(I,3),P(I,4),P(I,5),P(I,6)
```

```
280 READ Q(I,1),Q(I,2),Q(I,3),Q(I,4),Q(I,5),Q(I,6)
290 P(I,7)=P(I,3)-P(I,1)
300 P(I,8)=P(I,4)-P(I,2)
310 P(I,9)=SQR(P(I,7)↑2+P(I,8)↑2)
320 P(I,7)=P(I,7)/P(I,9)
330 P(I,8)=P(I,8)/P(I,9)
340 NEXT I
350 FOR I=1 TO N
360 READ B(I)
370 SB(I)=B(I)
380 NEXT I
385 REM ASSEMBLE STRUCTURE STIFFNESS MATRIX
390 FOR R=1 TO NM
400 GOSUB 10000
410 FOR I=1 TO 6
420 FOR J=1 TO 6
430 A(Q(R,I),Q(R,J))=A(Q(R,I),Q(R,J))+M(I,J)
440 NEXT J,I,R
445 REM SOLVE EQUATIONS
450 GOSUB 20000
460 F$="99   S999.999   S999.9999999"
470 PRINT#3,F$
480 PRINT#1,"NO.    LOAD    DISPLACEMENT"
490 GOSUB 5000
495 REM PRINT DISPLACEMENTS
500 FOR R=1 TO N
510 PRINT#1
520 PRINT#2,R,SB(R),X(R)
530 NEXT R
540 GOSUB 5000
545 REM PRINT GLOBAL MEMBER FORCES
550 PRINT#1,"                 GLOBAL MEMBER FORCES"
560 GOSUB 5000
570 F$="99 S9999.999 S9999.999 S9999.999 S9999.999 S9999.999 S9999.999"
580 PRINT#3,F$
590 PRINT#1,"NO.         FXA        FYA         MA         FXB        FYB         MB"
600 GOSUB 5000
610 FOR R=1 TO NM
620 FOR I=1 TO 6
630 BB(I,1)=X(Q(R,I))
640 NEXT I
650 GOSUB 10000
660 FOR I=1 TO 6
670 FOR J=1 TO 6
680 AA(I,J)=M(I,J)
690 NEXT J,I
700 GOSUB 12000
710 FOR J=1 TO 6
720 GM(R,J)=CC(J,1)
730 NEXT J
740 GOSUB 16000
```

```
750 NEXT R
760 GOSUB 5000
765 REM PRINT LOCAL MEMBER FORCES
770 PRINT#1,"                 LOCAL MEMBER FORCES"
780 GOSUB 5000
790 PRINT#1,"NO.        PA        VA        MA        PB        VB        MB"
800 GOSUB 5000
810 FOR R=1 TO NM
820 FOR J=1 TO 6
830 BB(J,1)=GM(R,J)
840 NEXT J
850 GOSUB 14000
860 FOR I=1 TO 6
870 FOR J=1 TO 6
880 AA(I,J)=TM(I,J)
890 NEXT J,I
900 GOSUB 12000
1000 GOSUB 16000
1010 NEXT R
1020 GOSUB 5000
1030 GOTO 30000
5000 PRINT#1
5010 PRINT#1
5020 PRINT#1,D$
5030 PRINT#1
5040 PRINT#1
5050 RETURN
9000 REM SUBROUTINE TO CALCULATE MEMBER STIFFNESS MATRIX
10000 L=P(R,9)
10010 L2=L*L
10020 L3=L*L*L
10030 S=P(R,8)
10040 S2=S*S
10050 C=P(R,7)
10060 C2=C*C
10070 AE=P(R,5)*E
10080 EI=P(R,6)*E
10090 K1=AE*C2/L+12*EI*S2/L3
10100 K2=((AE/L)-(12*EI/L3))*S*C
10110 K3=AE*S2/L+12*EI*C2/L3
10120 K4=6*EI*S/L2
10130 K5=6*EI*C/L2
10140 K6=2*EI/L
10150 M(1,1)=K1
10160 M(4,4)=K1
10170 M(1,2)=K2
10180 M(4,5)=K2
10190 M(1,3)=-K4
10200 M(1,6)=-K4
10210 M(1,4)=-K1
10220 M(1,5)=-K2
10230 M(2,4)=-K2
10240 M(2,2)=K3
10250 M(5,5)=K3
10260 M(2,3)=K5
10270 M(2,6)=K5
10280 M(2,5)=-K3
10290 M(3,3)=2*K6
10300 M(6,6)=2*K6
10310 M(3,4)=K4
```

```
10320 M(4,6)=K4
10330 M(3,5)=-K5
10340 M(5,6)=-K5
10350 M(3,6)=K6
10360 FOR I=2 TO 6
10370 FOR J=1 TO I-1
10380 M(I,J)=M(J,I)
10390 NEXT J,I
10400 RETURN
11000 REM MATRIX MULTIPLICATION
12000 M=6
12010 N=6
12020 P=1
12030 FOR I=1 TO M
12040 FOR J=1 TO P
12050 CC(I,J)=0
12060 FOR K=1 TO N
12070 CC(I,J)=CC(I,J)+AA(I,K)*BB(K,J)
12080 NEXT K,J,I
12090 RETURN
13000 REM TRANSFORMATION MATRIX
14000 FOR I=1 TO 6
14010 FOR J=1 TO 6
14020 TM(I,J)=0
14030 NEXT J,I
14040 S=P(R,8)
14050 C=P(R,7)
14060 TM(1,1)=C
14070 TM(2,2)=C
14080 TM(4,4)=C
14090 TM(5,5)=C
14100 TM(3,3)=1
14110 TM(6,6)=1
14120 TM(1,2)=S
14130 TM(4,5)=S
14140 TM(2,1)=-S
14150 TM(5,4)=-S
14160 RETURN
16000 PRINT#2,R,CC(1,1),CC(2,1),CC(3,1),CC(4,1),CC(5,1),CC(6,1)
16010 PRINT#1
16020 RETURN
19000 REM SUBROUTINE TO SOLVE EQUATIONS
20000 FOR I=1 TO N-1
20010 FOR J=I+1 TO N
20020 D=A(J,I)/A(I,I)
20030 FOR K=1 TO N
20040 A(J,K)=A(J,K)-D*A(I,K)
20050 NEXT K
20060 B(J)=B(J)-D*B(I)
20070 NEXT J,I
20080 X(N)=B(N)/A(N,N)
20090 FOR I=1 TO N-1
20100 J=N-I
20110 D=0
20120 FOR K=J+1 TO N
20130 D=D+A(J,K)*X(K)
20140 NEXT K
20150 X(J)=B(J)/A(J,J)-D/A(J,J)
20160 NEXT I
20170 RETURN
30000 END
```

Output Example 7.2

```
FRAME ANALYSIS

MODULUS OF ELASTICITY= 1
NUMBER OF MEMBERS= 2
NUMBER OF UNKNOWN DISPLACEMENTS= 4

NO.    LOAD    DISPLACEMENT

 1   +   .000   +   .0000000
 2   -  1.000   -   .2970000
 3   +   .000   -   .1350000
 4   +   .000   +   .3150000
```

GLOBAL MEMBER FORCES

NO.	FXA	FYA	MA	FXB	FYB	MB
1 +	.000 +	.722 +	.800 +	.000 -	.722 +	.499
2 +	.000 -	.277 -	.500 +	.000 +	.277 +	.000

LOCAL MEMBER FORCES

NO.	PA	VA	MA	PB	VB	MB
1 +	.000 +	.722 +	.800 +	.000 -	.722 +	.499
2 +	.000 -	.277 -	.500 +	.000 +	.277 +	.000

Output Example 7.3

```
FRAME ANALYSIS

MODULUS OF ELASTICITY= 210000000
NUMBER OF MEMBERS= 7
NUMBER OF UNKNOWN DISPLACEMENTS= 15

NO.     LOAD    DISPLACEMENT

 1   +  3.000   +   .0034701
 2   +   .000   -   .0012674
 3   +   .000   +   .0004746
 4   +   .000   +   .0034618
 5   -  5.000   +   .0003982
 6   +   .000   +   .0007727
 7   +   .000   +   .0034602
 8   +   .000   +   .0007715
 9   +   .000   +   .0006905
10   +  6.000   +   .0036133
11   +   .000   -   .0013269
12   +   .000   -   .0001113
13   +   .000   +   .0035917
14   +   .000   +   .0008422
15   +   .000   -   .0001529
```

GLOBAL MEMBER FORCES

NO.	FXA	FYA	MA	FXB	FYB	MB
1	+ 2.247	− .635	− 1.720	− 2.247	+ .635	+ .132
2	+ 2.667	− 2.281	− 6.190	− 2.667	+ 2.281	− 6.355
3	− .752	− .635	+ .337	+ .752	+ .635	+ 1.720
4	+ 2.247	− 5.635	+ 2.685	− 2.247	+ 5.635	+ .667
5	− 4.085	− 2.916	+ 6.115	+ 4.085	+ 2.916	+ 5.852
6	+ 4.914	− 7.916	+ 5.687	− 4.914	+ 7.916	+ 6.053
7	+ 2.247	− 5.635	− .132	− 2.247	+ 5.635	− 2.685

LOCAL MEMBER FORCES

NO.	PA	VA	MA	PB	VB	MB
1	+ 2.247	− .635	− 1.720	− 2.247	+ .635	+ .132
2	+ 2.667	− 2.281	− 6.190	− 2.667	+ 2.281	− 6.355
3	− .859	+ .481	+ .337	+ .859	− .481	+ 1.720
4	+ 6.012	+ .813	+ 2.685	− 6.012	− .813	+ .667
5	− 4.165	+ 2.801	+ 6.115	+ 4.165	− 2.801	+ 5.852
6	+ 8.871	+ 2.847	+ 5.687	− 8.871	− 2.847	+ 6.053
7	+ 2.247	− 5.635	− .132	− 2.247	+ 5.635	− 2.685

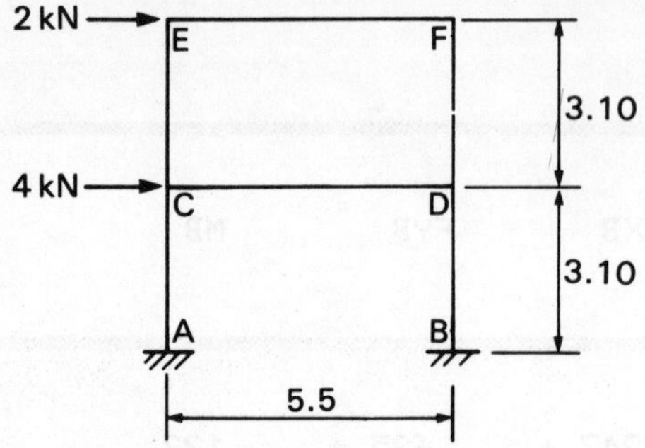

Fig. 7.10

7.5 Calculate the end moments and axial forces in both beams of the frame shown in Fig. 7.10. The properties are as follows:

Both beams: Area = 85.3 cm^2, I = 19 500 cm^4

All columns: Area = 58.8 cm^2, I = 4600 cm^4

7.6 For the four-span continuous beam shown in Fig. 7.11, obtain the influence line diagram for the bending moment at support B by analysing the structure repeatedly for a unit vertical load placed at every tenth point of each span in succession.

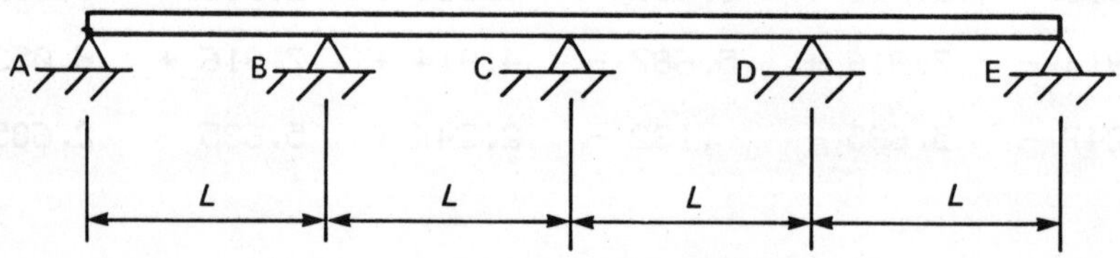

Fig. 7.11

Further Reading

1. Bhatt, P. *Problems in Structural Analysis by Matrix Methods*. Construction Press, 1981.
2. Ghali, A. and Neville, A. M. *Structural Analysis*. Chapman & Hall, 1978.
3. Jenkins, W. M. *Structural Mechanics and Analysis*. Nelson, 1982.

8

Applications in Structural Design

8.1 Introduction — Iterative Approach to Engineering Design

Engineering design is generally a process of selecting particular values of design parameters in such a way that the design requirements (constraints) are met in a practicable, economical way. The process is greatly assisted by the experience of the designer since this enables unlikely values of the parameters to be eliminated and trials made with likely values. An experienced designer will also be able to interpret the results of the trial designs and thus move more rapidly to a successful conclusion of the design process. The total design process is of course more complicated than this since it involves consideration of the fundamental design problem and its statement or description in engineering terms. For example, if we want to design a beam, we have already decided, perhaps tentatively, that a 'beam' will solve our engineering problem. We then go on to take our second fundamental decision, that the beam will be in reinforced concrete, and not in structural steel, for example.

The next stage of the design is the one we wish to concentrate on, for here we can transfer most of the work to the computer and develop a very rapid method of producing the design details. This will leave the designer free to concentrate on the important aspects of the design and *detach* himself from the tedious calculations and the constant reference to design tables which can dominate design carried out in the *manual* mode.

8.2 Design of Reinforced Concrete Beam to CP 110

As an illustration of an iterative design process we consider the design of a simple rectangular, singly reinforced, concrete beam to CP 110. We use the 'simplified' theory allowed by CP 110 and assume the design to be 'balanced' or 'under-reinforced'. In order to handle the 'self-weight' bending moment M_s, we assume the beam to be simply-supported and of uniform cross-section over span L. The applied bending moment for design purposes is M_A and this is taken to include *factored* dead and live loads but to exclude self-weight. Thus the design ultimate moment is

$$M_u = M_s + M_A \qquad (8.1)$$

The minimum effective depth is that corresponding to balanced design and is given by

$$d_{min} = \sqrt{\left(\frac{M_u}{0.15 f_{cu} b}\right)} \qquad (8.2)$$

where f_{cu} = concrete stress, and b = width of beam. If d is the chosen effective depth ($\geq d_{min}$), we can calculate the depth to the neutral axis x as

$$x = d - \sqrt{\left(d^2 - \frac{5 M_u}{f_{cu} b}\right)} \qquad (8.3)$$

Then we calculate the steel area required,

$$A_s = \frac{M_u}{0.87 f_y (d - x/2)} \qquad (8.4)$$

where f_y = steel stress. This is the theoretical steel area required, the actual area provided

Program 8.1

```
10 PRINT"****************************************"
15 PRINT"ITERATIVE DESIGN OF S R CONCRETE BEAM"
20 PRINT"TYPE IN APPLIED BENDING MOMENT (KNM)"
30 INPUT MA
34 PRINT"TYPE IN SPAN (METRES)"
36 INPUT L
40 PRINT"TYPE IN CONCRETE STRESS (N/SQMM)"
50 INPUT FC
60 PRINT"TYPE IN STEEL STRESS (N/SQMM)"
70 INPUT FY
80 PRINT"TYPE IN UNIT COST OF CONCRETE"
90 INPUT C
100 PRINT"TYPE IN UNIT COST OF STEEL"
110 INPUT S
120 PRINT"TYPE IN ESTIMATED WIDTH (MM)"
130 INPUT B
140 PRINT"TYPE IN ESTIMATED DEPTH(MM)"
150 INPUT D
160 PRINT"TYPE IN ESTIMATED COVER TO STEEL (MM)"
170 INPUT G
180 H=D+G
190 MS=2943*B*H*1E-9*L↑2*1.4
200 MU=MS+MA
210 PRINT"MU=";MU;"(KNM)"
215 MU=1E6*MU
220 DM=SQR(MU/(.15*FC*B))
230 PRINT"MINIMUM DEPTH =";DM
240 PRINT"DO YOU WISH TO REVISE B AND/OR D? Y/N"
250 GETA$:IF A$=""THEN250
260 IFA$="Y"THEN 120
300 X=D-SQR(D↑2-5*MU/(FC*B))
310 AS=MU/(.87*FY*(D-.5*X))
320 PRINT"MINIMUM STEEL AREA FOR THIS DESIGN =";AS
330 PRINT"TYPE IN CHOSEN STEEL AREA"
340 INPUT AS
350 CO=C*B*H*1E-6 + S*AS
360 PRINT"COST PER METRE OF THIS BEAM IS";CO
370 PRINT:PRINT"DO YOU WISH TO REVISE THE DESIGN? Y/N"
380 GETA$:IF A$=""THEN 380
390 IF A$="Y" THEN 120
395 PRINT"***************************************"
400 PRINT"DESIGN NOW COMPLETE"
410 PRINT"BREADTH=";B
420 PRINT"EFFECTIVE DEPTH=";D
430 PRINT"OVERALL DEPTH =";H
440 PRINT"STEEL AREA=";AS
450 PRINT"COST PER METRE =";CO
```

```
455 PRINT"***************************************"
460 PRINT"DO YOU WISH TO DESIGN ANOTHER BEAM? Y/N"
470 GETA$:IFA$=""THEN470
480 IFA$="Y"THEN10
490 END
```

Program 8.2

```
10 PRINT"PLASTIC DESIGN OF PITCHED ROOF PORTAL FRAME"
20 PRINT"FRAME HAS FIXED BASES"
30 PRINT"DESIGN IS FOR VERTICAL LOADING ONLY"
40 PRINT"TYPE IN W,L,H1,H2,A,N,FY"
50 INPUT W,L,H1,H2,A,N,FY
60 DIM D(28),B(28),M(28),A(28),Z(28)
70 FOR I=1 TO 28
80 READ D(I),B(I),M(I),A(I),Z(I)
90 NEXT I
100 DATA 914,419,388,494.5,17657,914,419,343,437.5,15474
101 DATA 914,305,289,368.8,12583,914,305,253,322.8,10947
102 DATA 914,305,201,256.4,8362,762,267,197,250.8,7167
103 DATA 762,267,173,220.5,6197,686,254,170,216.6,5624
104 DATA 686,254,152,193.8,4997
105 DATA 610,305,149,190.1,4572,610,229,140,178.4,4146
106 DATA 610,229,125,159.6,3677,533,210,122,155.8,3203
107 DATA 533,210,101,129.3,2620,457,191,89,113.9,2014
108 DATA 457,191,82,104.5,1833,457,152,82,104.5,1800
109 DATA 457,152,74,95,1622,457,152,67,85.4,1441
110 DATA 406,178,67,85.5,1346,406,178,60,76,1194
111 DATA 356,171,57,72.2,1009,356,171,51,64.6,894.9
112 DATA 356,127,39,49.4,653.6,305,127,37,47.5,540.5
113 DATA 305,102,33,40.8,479.9,254,146,31,40,395.6
114 DATA 254,102,25,32.2,305.6
115 MP=W*L↑2*(1-4*A↑2)/(16*(1+H2*(1-2*A)/H1))
120 Z=MP*N*1E3/FY
130 I=0
140 I=I+1:IF Z(I)-Z>0 THEN 140
150 I=I-1
160 PRINT"DESIGN SECTION IS";D(I);"BY";B(I);"BY";M(I);"UB"
170 LF=Z(I)*FY/MP*1E-3
180 PRINT"LOAD FACTOR =";LF
190 P=W*L*N/(2*A(I))
200 PRINT"MEAN AXIAL STRESS =";P
210 END
```

will depend on the number and diameter of bars selected and will be obtained from tables of steel reinforcement. The area provided must not be less than A_S as calculated from Equation 8.4.

The designer will be interested in the cost of the beam he has just designed so we shall include in our program a facility to determine the cost of the beam per metre length. The designer can then 'experiment' with his design parameters to produce an 'economical' design.

If C = cost of concrete (£ / m³), S = cost of steel (£ / mm²/m), then the cost of the beam (£/m) is:

$$\text{COST} = C \times B \times H \times 10^{-6} + S \times A_s \quad (8.5)$$

where B and H are the width and overall depth (mm) respectively, and $H = D + G$.

Program 8.1 will carry out the design in an interactive mode and can be tested using the following data:

f_{cu} = 40 N/mm²
f_y = 425 N/mm²
M_A = 648 kN m
span = 18 m
C = 31
S = 0.001 963

The following is a possible design:

B = 300
D = 900
G = 50
A_S = 3770
COST = 16.24

The 'cost' figures are illustrative only.

8.3 Design of Pitched Roof Portal Frame in Structural Steelwork

Here we illustrate a simple design approach in which we use the computer not only to carry out the design computations, but also to store the design data associated with the universal beams from which our design will be selected. The frame in question is the fixed base, pitched roof portal frame shown in Fig. 8.1a. We use the plastic method of design, which is very suitable for this structure, and design for purely vertical loading at a load factor of 1.7. The usual practice would be to design for dead load + imposed load at a load factor of 1.7 and then check the design for dead load and imposed load + wind load, at a load factor of 1.36.

The failure mechanism is shown in Fig. 8.1b. Plastic hinges develop at the column feet, at the eaves and also in the rafter on either side of the ridge. These latter plastic hinges are assumed to occur at the first purlin points down from the ridge. This is known to be a good approximation in practice.

By writing down equations for the plastic

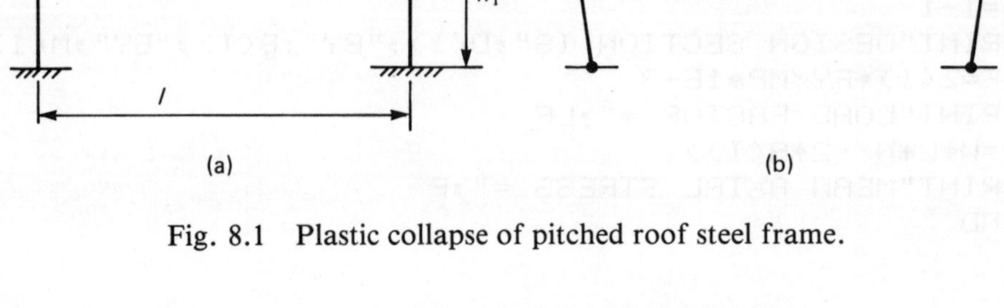

Fig. 8.1 Plastic collapse of pitched roof steel frame.

moment of resistance M_p, in terms of the frame geometry and factored loads, at the three plastic hinge positions in one symmetrical half of the structure and solving these equations we obtain

$$M_p = \frac{Nwl^2 (1 - 4\alpha^2)}{16\left[1 + \frac{h_2}{h_1}(1 - 2\alpha)\right]} \quad (8.6)$$

where w = intensity of dead load + imposed load, l = span, αl = purlin spacing, h_1 = height to eaves, h_2 = vertical distance from eaves to ridge, and N = load factor.

Having computed the M_p value required for the design, the next step is to compute the plastic section modulus required:

$$Z'_p = M_p/\sigma_y \quad (8.7)$$

where Z'_p = plastic section modulus required, and σ_y = yield stress.

A universal beam section is then to be selected from the list such that the plastic section modulus Z_p provided is not less than that required as determined from Equation 8.7.

The procedure adopted in the program is to store all the available sections (suitable for plastic design) in descending order of magnitude of Z_p; then scan the sections until (Z_p – Z'_p) is negative, then go back one section in the list and choose this section for the design. The actual load factor based on the choice of section is then computed and also the mean axial stanchion stress.

The universal beam data are stored section by section in the following order: D (depth), B (width, mass/metre), A (cross-sectional area, cm^2) and Z_p (cm^3). The program parameters are:

$W = w$
$L = l$
$H1 = h_1$
$H2 = h_2$
$A = \alpha$
N = Load factor
$FY = \sigma_y$

The member data are selected from the section book and are not comprehensive. Program 8.2 may be tested with the following set of data (taken from *Plastic Design*, Constrado, 1975).

W = 5.428 kN/m
L = 25 m
$H1$ = 7.6 m
$H2$ = 3.75 m
A = 0.05
N = 1.7
FY = 255 N/mm²

for which the program selects: 356 × 171UB57.

8.4 Interactive Computer-Aided Structural Design

The introduction of the computer into the structural design process has been slow. In the early days of the development, the benefits of the computer were almost entirely in the analysis of structures where the speed and accuracy of computers were welcomed. Progress was slow because computing was limited to 'batch' processing in which there is a significant delay in getting results from the computer. There were also problems arising from the handling of large quantities of data, which required checking, and output from the computer which needed skilled and painstaking interpretation. All this had to be carried out before the results of the analysis could have any influence on the actual design process. Clearly the analysis is only one step in the process and it is inconvenient to have to extract all the results from the computer simply to examine these and put critical values back in to continue the design.

Obviously it would be much more efficient to incorporate the analysis and the sorting of the results into a complete 'analysis and design' procedure. The difficulty in taking this step is that usually the designer has to apply judgement based on knowledge and experience in his interpretation of the results of the analysis, and this cannot be achieved satisfactorily in a batch-processing mode. The development of an 'interactive' computing mode was required to allow an integration of the various stages in the total design process.

In interactive computing, the designer can combine his knowledge, experience and judgement with the power and speed of the computer. Operating from a computer terminal, with printing and graphics facilities, the designer can engage in 'dialogue' with the computer in such a way that both designer and computer are used to best advantage. Moreover, the designer can work at his own pace since while the computer is waiting for his responses, it can be serving other users through 'time sharing'.

Once the design has been completed, the computer can take over the production of the working drawings. Here again the designer can use his experience and judgement to ensure that the conversion of the design into working details results in a practicable solution.

The developments just described have had a profound influence on structural design and have enabled designers to achieve more economical and more ambitious designs and more effective use of structural materials. Designers have always sought 'good', or even 'best', designs and experience and flair have enabled good designers to produce effective and economical designs. If it is possible to specify the criterion for the 'best' design, for example minimum cost, then it is now possible to use the computer to find the 'optimum' design.

The optimization of structural design is a subject in which considerable progress has been made in recent years. The future development of computer-aided design will certainly include procedures for optimizing design. A brief introduction to this topic was attempted in Section 5.7. The reader is recommended to study some of the selected literature on the subject.

Further Reading

Hung, E. J. and Arora, J. S. *Applied Optimal Design: Mechanical and Structural Systems*. Wiley, 1979.

Stark, R. M. and Nicholls, R. L. *Mathematical Foundations for Design: Civil Engineering Systems*. McGraw-Hill, 1972.

Svenson, N. L. *Introduction to Engineering Design*. Pitman, 1976.

9

Applications in Surveying

9.1 Calculations Relating to Closed Traverses

The main object here is to calculate the area of a closed traverse from data obtained in the field. This consist essentially of the lengths and bearings of the sides of the traverse. A convenient procedure often employed is the so-called 'Double Meridian Distance' (DMD) method. This is used as a basis for Program 9.1

9.1.1 Preliminary Calculations

The first step in calculating a closed traverse is to balance the angles. This means that the angles should be corrected or adjusted to conform with the geometric requirements of the figure. If any inconsistencies occur, then the angles are adjusted according to some generally accepted procedure. Details can be found in references listed at the end of the chapter to which the reader should refer. (It is assumed here that this adjustment has already been made and that all the angles of the figure fulfil the geometric requirements.)

9.1.2 Latitudes and Departures

The 'Latitude' of any line is its projection on a North–South line, and the 'Departure' is its projection on an East–West line. The determination of the latitudes and departures is the second stage of the calculation and their summations give an indication of the linear closing error of the traverse.

Where the sides of a polygon close, the algebraic sum of its latitudes and the algebraic sum of its departures are both equal to zero. It is implied here that the latitudes and departures have signs, and it is usual to adopt the following sign convention:

Latitudes North and Departures East are Positive.

It is only in very rare circumstances that the latitudes and departures of a traverse satisfy the requirements of closure. Usually, the summations of the latitudes and departures do not equal zero and this means that the polygon is in fact 'open'. The area of such an 'open' polygon, therefore, cannot be determined unless adjustments are made to close the figure.

9.1.3 Error Distribution

Before the area of a polygon can be determined, adjustments must be made to ensure that a mathematically closed figure is obtained. Many rules are available to effect the adjustment, each rule being suitable to a given set of known circumstances. If, for example, it is known that the error in the position of a point is due as much to the error in bearing as to the error in distance, then the 'Bowditch rule' may be applied. This rule states that the correction to be made to the latitude or departure of any line is to the total error in latitude or departure, as the length of the line is to the perimeter of the polygon.

This rule should be applied to the latitude and departure of each line resulting in quantities known as 'balanced latitudes' and 'balanced departures'. It will be found that after the corrections have been made, the

Program 9.1 Closed traverse calculations — program and output

```
1 DATA 10
2 DATA 103.75,-1,38,7,15,1
3 DATA 96.75,1,86,52,30,1
4 DATA 420.77,-1,39,18,30,1
5 DATA 208.64,1,62,31,30,1
6 DATA 436.79,1,25,56,30,-1
7 DATA 56.48,-1,87,1,15,-1
8 DATA 98.8,-1,53,22,0,-1
9 DATA 68.62,1,36,38,0,-1
10 DATA 95.1,1,59,29,0,-1
11 DATA 207.41,-1,51,40,45,-1
500 OPEN 1,4
510 OPEN 2,4,1
520 OPEN 3,4,2
530 F$="9999.999   S9   999  99  99  S9"
540 PRINT#3,F$
550 FOR I=1 TO 80
560 D$=D$+"▒"
570 NEXT I
580 GOSUB 5000
590 PRINT#1,"CLOSED TRAVERSE; DEG,MIN,SEC BEARINGS"
600 GOSUB 5000
610 PRINT#1,"AREA OF A CLOSED TRAVERSE BY DOUBLE MERIDIAN DISTANCE"
620 PRINT#1,"BOWDITCH CORRECTION APPLIED"
630 GOSUB 5000
640 READ NL
650 PRINT#1,"NO. OF SIDES =";NL
660 GOSUB 5000
670 PRINT#1,"DISTANCE              BEARING"
680 GOSUB 5000
690 DIM A(NL,11)
700 ES=0
710 EL=0
720 ED=0
730 FOR I=1 TO NL
735 REM READ DATA FOR EACH SIDE OF THE TRAVERSE
740 READ A(I,1),A(I,2),A(I,3),A(I,10),A(I,11),A(I,4)
750 PRINT#2,A(I,1),A(I,2),A(I,3),A(I,10),A(I,11),A(I,4)
760 A(I,3)=(A(I,3)+A(I,10)/60+A(I,11)/3600)*π/180
770 A(I,5)=A(I,1)*A(I,2)*COS(A(I,3))
780 A(I,6)=A(I,1)*A(I,4)*SIN(A(I,3))
790 ES=ES+A(I,1)
800 EL=EL+A(I,5)
810 ED=ED+A(I,6)
820 NEXT I
830 GOSUB 5000
840 DL=EL/ES
850 DD=ED/ES
860 SL=0
870 SD=0
875 REM APPLY CORRECTION TO LATITUDES AND DEPARTURES
880 FOR I=1 TO NL
890 A(I,7)=A(I,5)-A(I,1)*DL
900 A(I,8)=A(I,6)-A(I,1)*DD
910 SL=SL+A(I,7)
920 SD=SD+A(I,8)
930 NEXT I
935 REM COMPUTE DOUBLE MERIDIAN DISTANCES
940 A(1,9)=A(1,8)
```

```
950 FOR I=2 TO NL
960 A(I,9)=A(I-1,9)+A(I-1,8)+A(I,8)
970 NEXT I
980 DA=0
990 FOR I=1 TO NL
1000 DA=DA+A(I,9)*A(I,7)
1010 NEXT I
1020 GOSUB 5000
1030 PRINT#1,"LATITUDE     DEPARTURE     BAL. LAT.    BAL. DEP."
1040 GOSUB 5000
1050 F$="S9999.999    S9999.999    S9999.999    S9999.999"
1060 PRINT#3,F$
1070 FOR I=1 TO NL
1080 PRINT#1
1090 PRINT#2,A(I,5),A(I,6),A(I,7),A(I,8)
1100 NEXT I
1110 GOSUB 5000
1120 PRINT#2,EL,ED,SL,SD
1130 GOSUB 5000
1140 PRINT#1,"AREA =";DA/2;"SQUARE UNITS"
1150 GOSUB 5000
1160 GOTO 30000
5000 PRINT#1
5010 PRINT#1
5020 PRINT#1,D$
5030 PRINT#1
5040 PRINT#1
5050 RETURN
30000 END
```

CLOSED TRAVERSE; DEG,MIN,SEC BEARINGS

AREA OF A CLOSED TRAVERSE BY DOUBLE MERIDIAN DISTANCE
BOWDITCH CORRECTION APPLIED

NO. OF SIDES = 10

DISTANCE		BEARING			
103.750	-1	38	7	15	+1
96.750	+1	86	52	30	+1
420.770	-1	39	18	30	+1
208.640	+1	62	31	30	+1
436.790	+1	25	56	30	-1
56.480	-1	87	1	15	-1
98.800	-1	53	22	0	-1
68.620	+1	36	38	0	-1
95.100	+1	59	29	0	-1
207.410	-1	51	40	45	-1

LATITUDE	DEPARTURE	BAL. LAT.	BAL. DEP.
- 81.621	+ 64.047	- 81.620	+ 64.049
+ 5.274	+ 96.606	+ 5.275	+ 96.608
- 325.569	+ 266.555	- 325.565	+ 266.565
+ 96.258	+ 185.107	+ 96.260	+ 185.113
+ 392.778	- 191.076	+ 392.783	- 191.065
- 2.935	- 56.403	- 2.934	- 56.402
- 58.953	- 79.284	- 58.952	- 79.281
+ 55.065	- 40.944	+ 55.066	- 40.943
+ 48.290	- 81.926	+ 48.291	- 81.924
- 128.607	- 162.723	- 128.605	- 162.718
- .019	- .043	- .000	+ .000

AREA = 148953.585 SQUARE UNITS

summations of both latitudes and departures equal zero and a closed polygon results. The calculation of the area can then be made using the double meridian distances.

9.1.4 *Double Meridian Distances*

Detailed descriptions of the method of calculating areas by double meridian distance are outside the scope of the present work and only the main features will be mentioned here.

1. A vertical reference line is drawn through the most westerly point of the traverse.
2. Going round each line of the traverse:
   (a) The DMD of the first line is equal to the departure of the line itself.
   (b) The DMD of any subsequent line equals the sum of the DMD and departure of the preceding line and the departure of the line itself.
3. The double area of the polygon equals the sum of the products of the DMD of a line and its own latitude; the summation being taken for all lines in the traverse.

In all calculations, due regard must be paid to the signs of all quantities involved. The double area mentioned in (3) above is the algebraic sum of all positive and negative area quantities. This should then be divided by two to obtain the final numerical value.

9.1.5 *Closed Traverse Program*

A listing of Program 9.1 to calculate the area of a closed polygon by the DMD method is provided. Some explanatory notes are included on the listing to aid understanding.

9.1.6 *Example of Use of Closed Traverse Program*

Table 9.1 and Fig. 9.1 show the basic information obtained from a field survey. From these, data preparation can proceed immediately since the program will apply the corrections according to the Bowditch rule automatically.

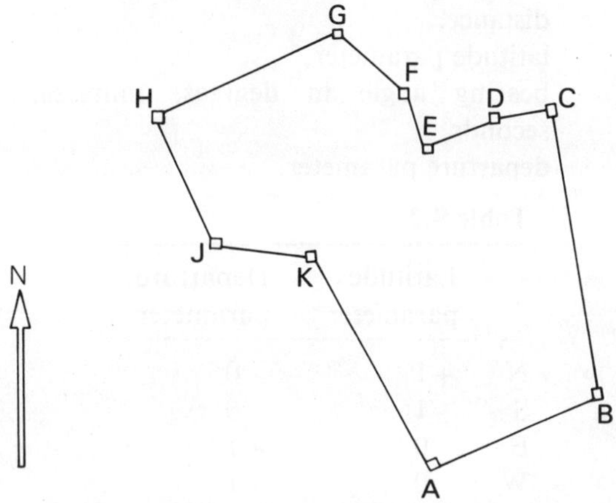

Fig. 9.1 A closed traverse.

9.1.7 *Latitude and Departure Parameters*

If the correct algebraic quantities are to be obtained, the bearing angles of each line of the traverse have to be described properly so that they produce latitudes and departures with the correct mathematical signs. Accordingly, the direction components (N,S,E,W) of the bearing angle are assigned the values in Table 9.2. With this notation, minimal programming effort is required and the preparation of data proceeds as follows:

Number of sides of the traverse:
For each side, its

Table 9.1 Distances and Bearing Angles

Line	Distance	Bearing
H J	103.75	S 38-07-15 E
J K	96.75	N 86-52-30 E
K A	420.77	S 39-18-30 E
A B	208.64	N 62-31-30 E
B C	436.79	N 25-56-30 W
C D	56.48	S 87-01-15 W
D E	98.80	S 53-22-00 W
E F	68.62	N 36-38-00 W
F G	95.10	N 59-29-00 W
G H	207.41	S 51-40-45 W

distance,
latitude parameter,
bearing angle in degrees, minutes, seconds,
departure parameter.

Table 9.2

	Latitude parameter	Departure parameter
N	+1	0
S	−1	0
E	0	+1
W	0	−1

9.1.8 Output from Closed Traverse Program

The data as presented to the program are also part of the output so that a check can be made if the correct entries have been made. Although not strictly necessary, information about the balanced latitudes and departures is also included in the output, again to provide a check on the calculated values. Finally, the computed area is printed out, correct to within three decimal places.

9.2 Earthwork Calculations

One of the more common computational problems in surveying is the determination of volumes of earthwork excavations. It is often necessary to compute not only the volumes, but also the location of the centres of gravity for estimation of haul distances. This problem is often encountered in highway engineering whenever roadway ground profiles have to be aligned.

Consideration will now be given to the situation where a field survey has already been carried out to determine the ground elevations at all stations along the centre line of a proposed roadway (see Fig. 9.2).

At each station along the line, it is assumed that sufficient data have been gathered to establish a '3 level section'. Thus, in Fig. 9.3, distances C, D_L, D_R, H_L, and H_R are known. These, together with the base width b and the distance L between stations which are assumed constants, will be sufficient to calculate the total volume of earthwork between the first and last station.

9.2.1 Basis of the Method

Fig. 9.4 shows a solid whose end faces are parallel and at a distance L from each other. Using the notation: b_0 = base of triangle ABC; h_0 = altitude of triangle ABC; b_1 = base of triangle DEF; h_1 = altitude of triangle

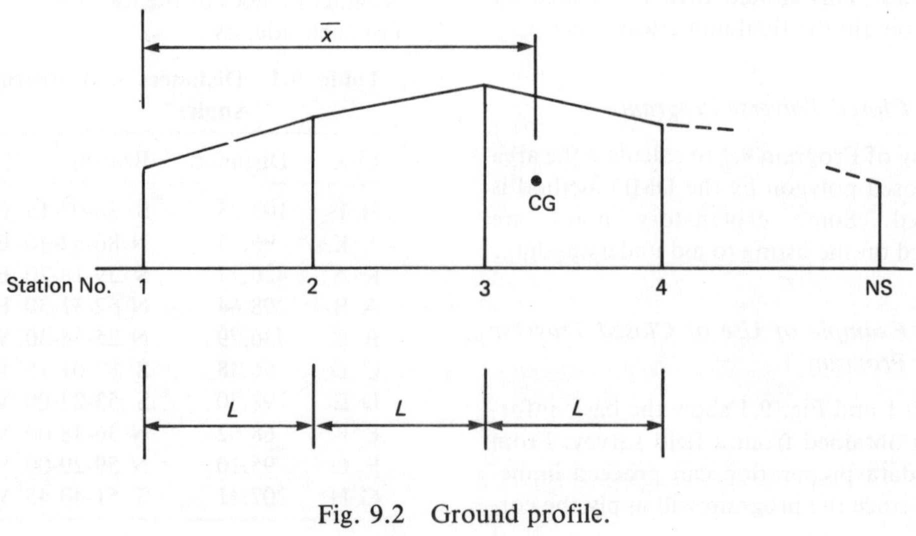

Fig. 9.2 Ground profile.

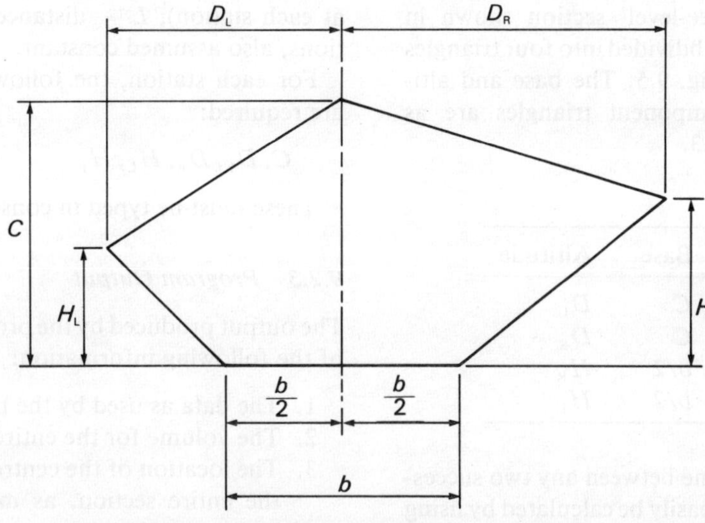

Fig. 9.3 Cross-section at each station.

DEF, then the volume of this solid can be shown to be

$$V = \frac{L}{12}\ (2\,b_0h_0 + 2\,b_1h_1 + b_0h_1 + b_1h_0) \qquad (9.1)$$

and its centroid is located at a distance $\bar{x}$ from the face ABC, where

$$\bar{x} = \frac{L^2}{24\,V}\ (b_0h_0 + b_0h_1 + b_1h_0 + 3\,b_1h_1) \qquad (9.2)$$

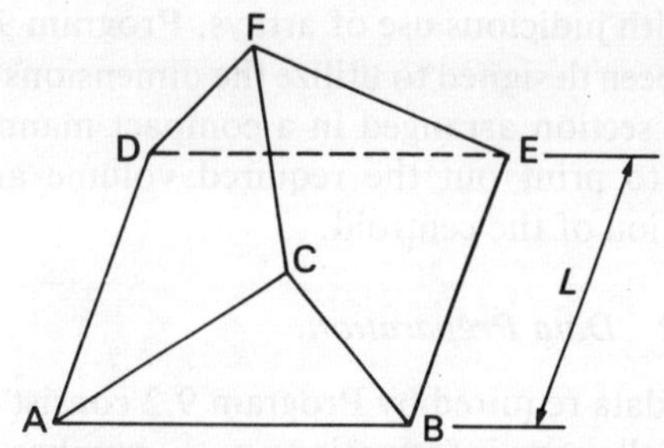

Fig. 9.4 Solid bounded by parallel triangular planes at ends.

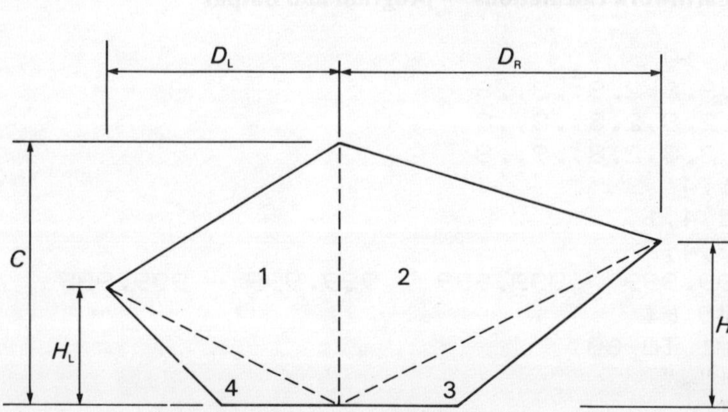

Fig. 9.5 Subdivision of cross-section into four triangles.

Now, the 'three-level' section shown in Fig. 9.3 may be subdivided into four triangles as indicated in Fig. 9.5. The base and altitudes of the component triangles are as shown in Table 9.3.

Table 9.3

Triangle	Base	Altitude
1	C	D_L
2	C	D_R
3	$b/2$	H_R
4	$b/2$	H_L

Thus, the volume between any two successive stations may easily be calculated by using Equation 9.1 four times. It is then just a simple matter to extend the calculation to cover the whole section.

With judicious use of arrays, Program 9.2 has been designed to utilize the dimensions at each section arranged in a compact manner and to print out the required volume and location of the centroid.

9.2.2 Data Preparation

The data required by Program 9.2 consist of the following information: n_S = number of stations; b = base width (assumed constant at each station); L = distance between stations, also assumed constant.

For each station, the following distances are required:

$$C, D_L, D_R, H_L, H_R$$

These must be typed in consecutively.

9.2.3 Program Output

The output produced by the program consists of the following information:

1. The data as used by the program.
2. The volume for the entire section.
3. The location of the centre of gravity of the entire section, as measured from station number 1.

A sample output from the program is provided following the program listing.

Further Reading

1. Clark, D. *Plane and Geodetic Surveying for Engineers*, Vol. 1, *Plane Surveying*. Metric edition. Constable, 1972.
2. Olliver, J. and Clendinning, J. *Principles of Surveying*, 4th edn., 2 vols. Van Nostrand Reinhold, 1978.

Program 9.2 Earthwork calculations — program and output

```
1 DATA 3,2,10
2 DATA .8,2.8,2.8,.8,.8
3 DATA .8,2.8,2.8,.8,.8
4 DATA .8,2.8,2.8,.8,.8
500 OPEN 1,4
510 OPEN 2,4,1
520 OPEN 3,4,2
530 F$="999.999   999.999   999.999   999.999   999.999"
540 PRINT#3,F$
550 FOR I=1 TO 80
560 D$=D$+"▒"
570 NEXT I
580 GOSUB 5000
590 PRINT#1,"EARTHWORKS CALCULATIONS"
600 GOSUB 5000
610 READ NS,B,L
```

```
620 PRINT#1,"NO. OF STATIONS =";NS
630 PRINT#1,"BASE WIDTH =";B
640 PRINT#1,"DISTANCE BETWEEN STATIONS =";L
650 GOSUB 5000
660 DIM A(5,NS),B(10,NS-1)
670 FOR J=1 TO NS
680 READ A(1,J),A(2,J),A(3,J),A(4,J),A(5,J)
690 PRINT#1
700 PRINT#2,A(1,J),A(2,J),A(3,J),A(4,J),A(5,J)
710 NEXT J
720 FOR J =1 TO NS-1
730 B0=A(1,J)
740 H0=A(2,J)
750 B1=A(1,J+1)
760 H1=A(2,J+1)
770 GOSUB 10000
780 B(1,J)=V
790 B(5,J)=XC
800 B0=A(1,J)
810 H0=A(3,J)
820 B1=A(1,J+1)
830 H1=A(3,J+1)
840 GOSUB 10000
850 B(2,J)=V
860 B(6,J)=XC
870 B0=B/2
880 H0=A(5,J)
890 B1=B0
900 H1=A(5,J+1)
910 GOSUB 10000
920 B(3,J)=V
930 B(7,J)=XC
940 H0=A(4,J)
950 H1=A(4,J+1)
960 GOSUB 10000
970 B(4,J)=V
980 B(8,J)=XC
990 NEXT J
1000 FOR J=1 TO NS-1
1010 B(9,J)=0
1020 B(10,J)=0
1030 FOR I=1 TO 4
1040 B(10,J)=B(10,J)+B(I,J)*B(I+4,J)
1050 B(9,J)=B(9,J)+B(I,J)
1060 NEXT I
1070 B(10,J)=B(10,J)/B(9,J)
1080 NEXT J
1090 V=0
1100 VX=0
1110 FOR J=1 TO NS-1
```

```
1120 V=V+B(9,J)
1130 VX=VX+B(9,J)*(B(10,J)+(J-1)*L)
1140 NEXT J
1150 VX=VX/V
1160 GOSUB 5000
1170 PRINT#1,"TOTAL VOLUME OF EARTHWORK =";V;"CUBIC UNITS"
1180 GOSUB 5000
1190 PRINT#1,"C.G. OF VOLUME IS AT";VX;"LINEAR UNITS FROM STA.NO. 1"
1195 GOSUB 5000
1200 GOTO 30000
5000 PRINT#1
5010 PRINT#1
5020 PRINT#1,D$
5030 PRINT#1
5040 PRINT#1
5050 RETURN
10000 V=L*(2*B0*H0+2*B1*H1+B0*H1+B1*H0)/12
10010 XC=L*L*(B0*H0+B0*H1+B1*H0+3*B1*H1)/(24*V)
10020 RETURN
30000 END
```

```
EARTHWORKS CALCULATIONS
```

```
NO. OF STATIONS = 3
BASE WIDTH = 2
DISTANCE BETWEEN STATIONS = 10
```

```
  .800     2.800     2.800      .800      .800

  .800     2.800     2.800      .800      .800

  .800     2.800     2.800      .800      .800
```

```
TOTAL VOLUME OF EARTHWORK = 60.8 CUBIC UNITS
```

```
C.G. OF VOLUME IS AT 10 LINEAR UNITS FROM STA.NO. 1
```

10

Applications in Hydraulics

10.1 Introduction

Among the several computational problems that arise in the field of hydraulics, two cases have been selected to serve as further demonstration of the computer's ability to perform repetitive calculations.

The first of these is the determination of gradually varied flow profiles in open channels and the second is the analysis of flow in pipe networks. It is not the intention here to develop fully comprehensive and powerful programs but simply to show that in these and similar situations, the use of computers would simplify the task tremendously and cut down the time required to obtain results.

10.2 Gradually Varied Flow Profiles in Open Channels

Basically the problem is the determination of the water surface elevations, i.e. the profile of the stream under non-uniform flow conditions in open channels. A full discussion of the conditions surrounding this problem is outside the scope of this text and only an outline will be presented.

The determination of the water surface profile may be carried out by any one of a number of different methods but attention here will be directed to a so-called 'step' method. The idea is to divide the channel into a number of short intervals and to work out the surface elevations at the extremities of each interval.

The number of intervals into which the channel is to be divided is a variable, and the accuracy of the results will increase as the number of intervals increases. The normal depth of flow is needed for this problem and it will be assumed that this quantity can be evaluated beforehand. One can then divide the total change in flow depth into a suitable integer number, thus determining the depths of various sections along the channel length. The problem then reduces to one of locating the *positions* of these sections. Alternatively, it might well be that the depths do not have to be incremented equally, i.e. the problem being to find the location of sections with given depths. In this case, one can just assign the desired values of the flow depths as data to simplify the writing of programs.

Fig. 10.1 shows in elevation a portion of an open channel under non-uniform flow conditions. With the notation as indicated in the figure and as defined subsequently, the equality of the total heads at Sections 1 and 2 may be written down as

$$S_0 \,\Delta x + y_1 + \alpha_1 \frac{V_1^2}{2g} = y_2 + \alpha_2 \frac{V_2^2}{2g} + S_f \,\Delta x \qquad (10.1)$$

From which Δx may be solved for

$$\Delta x = \frac{\Delta E}{S_0 - S_f} \qquad (10.2)$$

where

$$\Delta E = \left[y_2 + \alpha_2 \frac{V_2^2}{2g} \right] - \left[y_1 + \alpha_1 \frac{V_1^2}{2g} \right] \qquad (10.3)$$

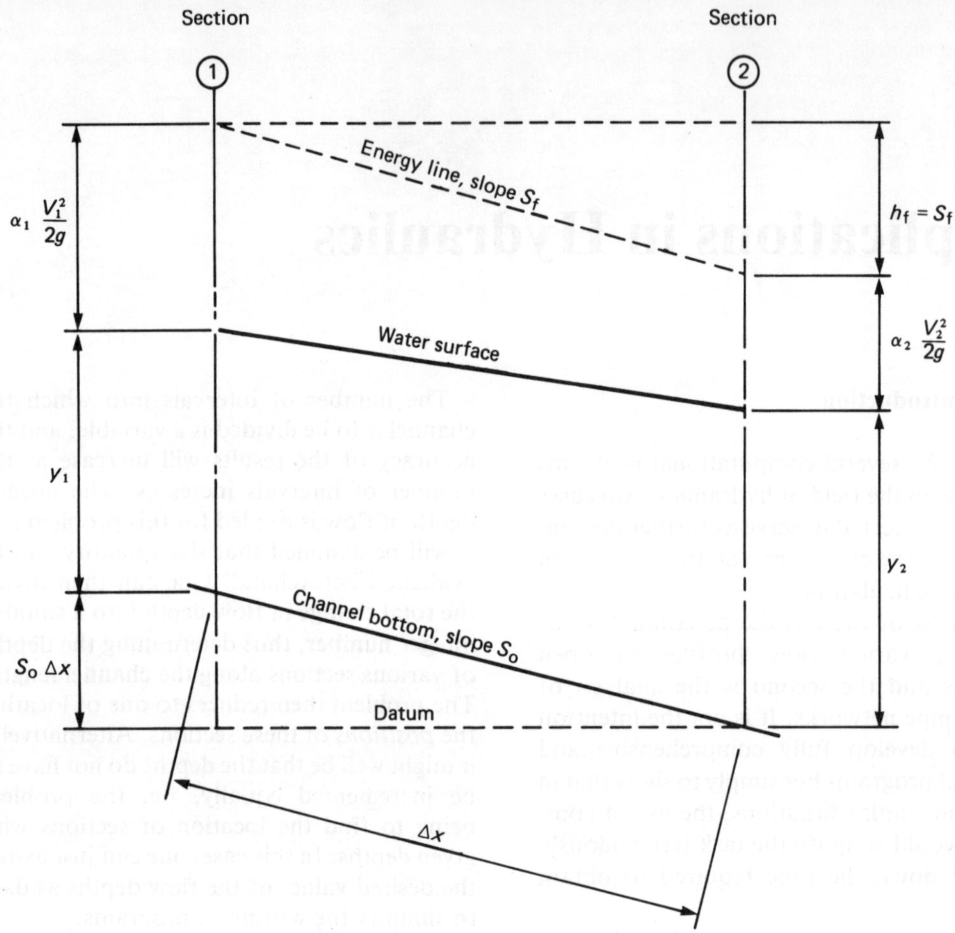

Fig. 10.1 Open channel with non-uniform flow.

The various terms are defined as follows:

S_0 = slope of channel bottom,
Δx = length of the channel between Sections 1 and 2,
y_1, y_2 = depths of water at Sections 1 and 2 respectively,
α_1, α_2 = energy coefficients for velocity heads to correct for the overall effect of non-uniform velocity distribution at a section (usually assumed = 1.0)
V_1, V_2 = mean velocities at Sections 1 and 2
S_f = friction slope (Manning)
= $n^2 V^2/R^{4/3}$ (metric units)
R = hydraulic radius

Equation 10.2 forms the basis of the solution as this yields the distance Δx in terms of the change in specific energy and the slope difference $(S_0 - S_f)$. This equation is applied over each interval bounded by two sections, one of which can be the reference section. Thus, by repeated application of the equation, the position of each succeeding section may be determined, and it will then be a simple matter to tabulate the cumulative distances from the starting position. This will make plotting of the results easier.

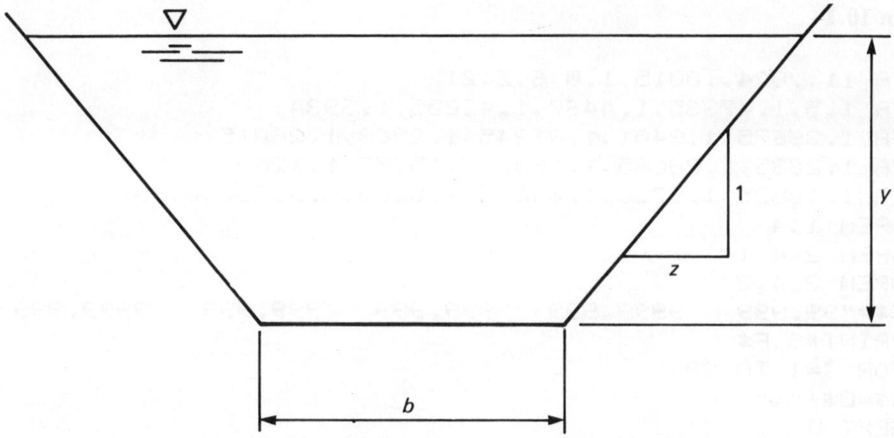

Fig. 10.2 Trapezoidal open channel section.

In Example 10.1, the variable S_f will be taken as the mean value over each interval, but the channel slope will be considered constant.

Example 10.1

A trapezoidal channel as in Fig. 10.2, with the following properties, carries a discharge of 11.0 m³/s. Base width = 6.0 m, side slope = 1.75, longitudinal slope = 0.0015, n (Manning's equation) = 0.024.

It is required to establish the flow profile created by a dam which backs up the water to a depth of 1.5 metres immediately behind the dam. Assume that the energy coefficient α = 1.0 and the normal depth = 0.967 metre.

The total change in flow depth = 1.5 − 0.967 = 0.533 metre and if this is divided into 20, say, the positions of 21 sections will be determined by using Program 10.1.

10.2.1 The Program

Program 10.1 has been written to provide the solution to problems involving a channel with trapezoidal section. This can, however, be altered easily to cater for different shapes of cross-sections by using the appropriate expressions for the cross-sectional area and hydraulic radius. The depths of flow at the different sections can be computer generated if desired, but this is left to the reader as an exercise. In the form presented here, the depths are given as part of the data.

The following is a complete specification of the data required by the program:

- Discharge in m³/s
- Friction factor in Manning's formula
- Channel bottom slope
- Energy coefficient
- Channel width and side slope
- Number of stations whose locations are to be determined
- Depths of flow at each station

The data are provided in 'data' statements and these may be located at suitable points in the program. Alternatively, if desired, the user may make the alterations so that the data are typed in from the keyboard as required by using 'input' statements.

The output follows the listing of Program 10.1. The three intermediate columns (area, velocity and energy) are not strictly required, but have been included here as a matter of interest to show the variation of these quantities along the channel length.

A plot of the results is shown in Fig. 10.3.

10.3 Flow in Pipe Networks

The problem of the analysis of flow in pipe networks lies in the determination of the dis-

Program 10.1

```
1 DATA 11,.024,.0015,1.0,6,2,21
2 DATA 1.5,1.47335,1.4467,1.42005,1.3934
3 DATA 1.36675,1.3401,1.31345,1.2868,1.26015
4 DATA 1.2335,1.20685,1.1802,1.15355,1.1269
5 DATA 1.10025,1.0736,1.04695,1.0203,0.99367,0.967
500 OPEN 1,4
510 OPEN 2,4,1
520 OPEN 3,4,2
530 F$="99.999   9999.999   999.999   999.999   9999.999"
540 PRINT#3,F$
550 FOR J=1 TO 80
560 D$=D$+"▒"
570 NEXT J
580 GOSUB 5000
590 PRINT#1,"NON-UNIFORM FLOW IN OPEN CHANNELS"
600 GOSUB 5000
610 READ Q,N,SO,AL,B,Z,M
620 DIM P(M,9)
630 FOR I=1 TO M
640 READ P(I,1)
650 NEXT I
660 G=9.81
670 Y=P(1,1)
680 GOSUB 1000
690 P(1,2)=A
700 V=Q/P(1,2)
710 GOSUB 2000
720 P(1,4)=E
730 GOSUB 3000
740 P(1,6)=SF
750 FOR I=2 TO M
760 Y=P(I,1)
770 GOSUB 1000
780 P(I,2)=A
790 V=Q/P(I,2)
800 GOSUB 2000
810 P(I,4)=E
820 P(I,5)=P(I,4)-P(I-1,4)
830 P(I,5)=ABS(P(I,5))
840 GOSUB 3000
850 P(I,6)=SF
860 P(I,7)=(P(I,6)+P(I-1,6))/2
870 P(I,8)=P(I,5)/ABS((SO-P(I,7)))
880 P(I,9)=P(I-1,9)+P(I,8)
890 NEXT I
900 PRINT#2,"DEPTH     AREA     VELOCITY     ENERGY     DISTANCE X"
910 GOSUB 5000
920 FOR I=1 TO M
930 PRINT#2,P(I,1),P(I,2),Q/P(I,2),P(I,4),P(I,9)
940 NEXT I
950 GOSUB 5000
960 GOTO 20000
```

```
1000 A=(B+Z*Y)*Y
1010 R=A/(B+2*Y*SQR(1+Z*Z))
1020 RETURN
2000 E=AL*V*V/(2*G)+Y
2010 RETURN
3000 SF=(N*V)↑2/(R↑(4/3))
3010 RETURN
5000 PRINT#1
5010 PRINT#1
5020 PRINT#1,D$
5030 PRINT#1
5040 PRINT#1
5050 RETURN
20000 END
```

NON-UNIFORM FLOW IN OPEN CHANNELS

DEPTH	AREA	VELOCITY	ENERGY	DISTANCE X
1.500	13.500	.814	1.533	.000
1.473	13.181	.834	1.508	22.023
1.446	12.866	.854	1.483	44.455
1.420	12.553	.876	1.459	67.355
1.393	12.243	.898	1.434	90.793
1.366	11.936	.921	1.410	114.856
1.340	11.632	.945	1.385	139.650
1.313	11.331	.970	1.361	165.310
1.286	11.032	.997	1.337	192.007
1.260	10.736	1.024	1.313	219.968
1.233	10.444	1.053	1.290	249.495
1.206	10.154	1.083	1.266	281.005
1.180	9.866	1.114	1.243	315.104
1.153	9.582	1.147	1.220	352.706
1.126	9.301	1.182	1.198	395.294
1.100	9.022	1.219	1.176	445.514
1.073	8.746	1.257	1.154	508.808
1.046	8.473	1.298	1.132	599.493
1.020	8.203	1.340	1.111	783.055
.993	7.936	1.385	1.091	1669.041
.967	7.672	1.433	1.071	1780.725

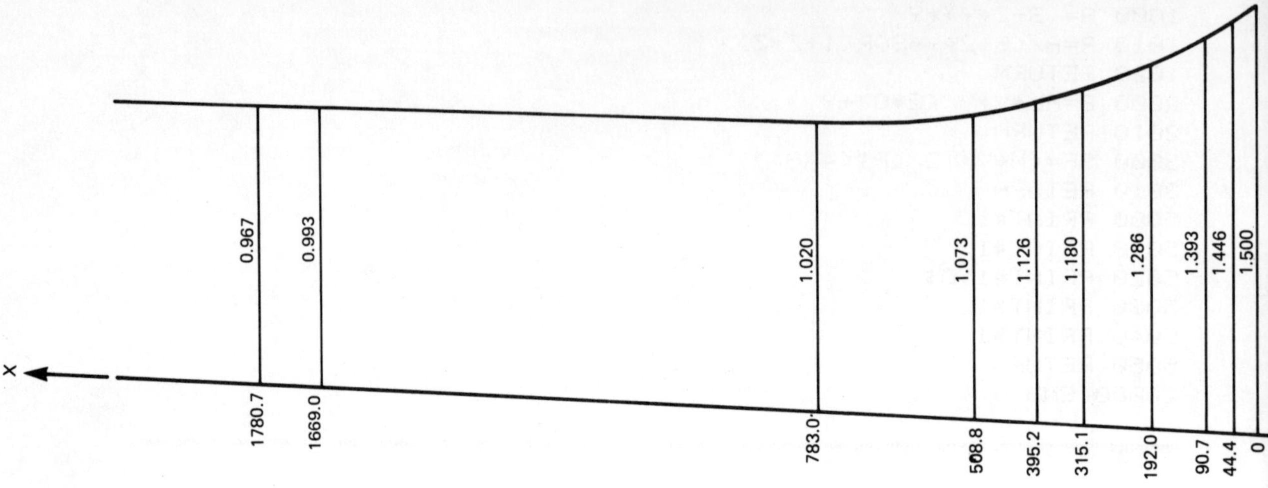

Fig. 10.3 Channel flow profile.

charge in each individual pipe of the system, The pipes are joined together at nodes where there could be additional inflow from, or outflow into, external sources.

The two conditions which must be satisfied in the analysis of any network problem are:

(a) The total discharge into a node must be equal to the total discharge out of that node; and
(b) in any closed loop, the algebraic sum of the head losses must equal zero.

Usually, only the loss of head due to friction is considered in pipe flow problems, and in this approach, other factors will not be taken into account.

The procedure adopted is based on the solution of a set of linear equations involving junction heads as the unknowns. This method is one of a number of possible approaches to the solution of the problem. Although convergence difficulties may arise for certain configurations, in most cases satisfactory solutions are usually obtained.

10.3.1 Theoretical Considerations

The loss of head in a pipe due to friction may be expressed by the equation

$$h_f = \left[\frac{32 f L}{\pi^2 g D^5}\right] Q^2 \qquad (10.4)$$

where f = friction coefficient, L, D = length and diameter of the pipe respectively, Q = discharge through the pipe, and g = acceleration due to gravity.

Let H_i and H_j be the respective heads at joints i and j to which a pipe is connected. Then

$$h_f = (H_i - H_j) = \frac{1}{C_{ij}} Q_{ij}^2 \qquad (10.5)$$

where

$$C_{ij} = \frac{\pi^2 g D_{ij}^5}{32 f_{ij} L_{ij}} \qquad (10.6)$$

The discharge may now be expressed in terms of the junction heads

$$Q_{ij}^2 = C_{ij}(H_i - H_j) \tag{10.7}$$

At any junction, several pipes may be joined together and the continuity of flow requires that

Total inflow = Total outflow

i.e.

$$\sum_1^p Q_{ij} = Q_{oj} \tag{10.8}$$

where P = number of pipes connected to the junction. Thus

$$\sum_1^p \frac{C_{ij}}{Q_{ij}} (H_i - H_j) = Q_{oj} \tag{10.9}$$

Equation 10.9 may be written for all junctions in the network resulting in a set of simultaneous equations involving junction heads. These equations also contain the discharges in each pipe and as these are not known at the beginning, it becomes necessary to assign arbitrary, initial values so that the solution may proceed. After the first cycle of calculations has been completed, the discharge in each individual pipe may now be recomputed, and these revised values substituted back again. This operation produces a new set of equations which are then solved. The intermediate values of discharges may be printed out if desired.

It is apparent, therefore, that the solution is an *iterative* one, and the calculations may be terminated as soon as convergence towards a steady solution is detected.

A slight difficulty arises in that the number of iterations required to produce convergence is not easy to deduce beforehand, although it may not be large. With this in view the program may be designed so that this may be fed in as part of the data in an interactive manner.

As already pointed out, each iteration depends on previously calculated (or assumed) values of the pipe discharges and Equation 10.9 may be rewritten as follows:

$$\Sigma \left(\frac{C_{ij}}{Q_{ij}}\right)^{n-1} (H_i - H_j)^n = Q_{oj} \tag{10.10}$$

with the superscripts indicating the number of the iteration.

For the complete network, the equations may be written in more compact matrix notation thus:

$$[K]^{n-1}\{H\}^n = \{Q_0\} \tag{10.11}$$

10.3.2 The Program

Program 10.2 has been written to solve for the unknown junction heads and evaluate the discharge and head loss in each pipe of the network. Provision is made to allow the assignment of particular values of friction coefficient to each pipe (see example). It is left to the user to decide the most appropriate values to apply in each case.

The program will invite the user to type in the desired number of iterations at the beginning of each run. Low values should be used initially. As mentioned before, some configurations may give rise to convergence problems, in which cases, alternative methods of solution will have to be used.

Matrix K of Equation 10.11 is formed automatically in statement numbers 520–600. Subroutine statements beginning at 11000 can be incorporated here if desired. The equations are solved by using the same subroutine as the one employed in the structural analysis programs of Chapter 7.

Program 10.2

```
10 DATA 12,9
11 DATA 182,.3,.05,.006,1,2
12 DATA 182,.3,.063,.006,2,3
13 DATA 182,.25,.032,.007,4,5
14 DATA 182,.25,.032,.007,6,5
15 DATA 182,.3,.032,.006,7,8
16 DATA 182,.3,.003,.006,8,9
17 DATA 122,.2,.063,.008,1,4
18 DATA 122,.25,.063,.007,2,5
19 DATA 122,.2,.05,.008,3,6
20 DATA 122,.2,.031,.008,4,7
21 DATA 122,.25,.031,.007,8,5
22 DATA 122,.2,.02,.008,6,9
23 DATA .1896,0,-.0158,0
24 DATA -0.158,0,0,0,-0.0158
200 OPEN 1,4
210 OPEN 2,4,1
220 OPEN 3,4,2
230 FOR I=1 TO 80
240 D$=D$+"▒"
250 NEXT I
260 F$="99  9999.999   9999.999  S99999.999  S99999.999"
270 PRINT#3,F$
280 GOSUB 5000
290 PRINT#1,"FLOW IN PIPE NETWORKS,DARCY-WEISBACH VERSION"
300 GOSUB 5000
310 READ NP,N
320 G=9.81
330 INPUT "NO. OF ITERATIONS REQ'D=";NI
340 PRINT#1,"NO. OF ITERATIONS=";NI
350 GOSUB 5000
360 PRINT#1,"PIPE   LENGTH   DIAMETER  DISCHARGE  HEAD LOSS"
370 GOSUB 5000
380 DIM P(NP,6),A(N,N),B(N),X(N),M(2,2),SN(NP,2),QE(N)
390 FOR I=1 TO NP
400 READ P(I,1),P(I,2),P(I,4),P(I,6),SN(I,1),SN(I,2)
410 P(I,3)=π*π*G*P(I,2)↑5/(32*P(I,6)*P(I,1))
420 NEXT I
430 FOR I=1 TO N
440 READ B(I)
450 QE(I)=B(I)
460 NEXT I
470 FOR IT=1 TO NI
480 FOR I=1 TO N
490 FOR J=1 TO N
500 A(I,J)=0
510 NEXT J,I
```

```
520 FOR R=1 TO NP
530 GOSUB 10000
540 GOSUB 11000
550 FOR I=1 TO 2
560 K=SN(R,I)
570 FOR J=1 TO 2
580 L=SN(R,J)
590 A(K,L)=A(K,L)+M(I,J)
600 NEXT J,I,R
610 FOR I=1 TO N
620 B(I)=QE(I)
630 NEXT I
635 GOSUB 20000
640 FOR R=1 TO NP
650 GOSUB 10000
660 P(R,4)=(K*(X(SN(R,1))-X(SN(R,2)))+P(R,4))/2
670 P(R,5)=P(R,4)↑2/P(R,3)
680 NEXT R
690 NEXT IT
700 FOR R=1 TO NP
710 PRINT#2,R,P(R,1),P(R,2),P(R,4),P(R,5)
720 NEXT R
730 GOSUB 5000
740 GOTO 30000
5000 PRINT#1
5010 PRINT#1
5020 PRINT#1,D$
5030 PRINT#1
5040 PRINT#1
5050 RETURN
10000 K=P(R,3)/ABS(P(R,4))
10010 RETURN
11000 M(1,1)=K
11010 M(2,2)=K
11020 M(1,2)=-K
11030 M(2,1)=-K
11040 RETURN
20000 FOR I=1 TO N-1
20010 FOR J=I+1 TO N
20020 D=A(J,I)/A(I,I)
20030 FOR K = 1 TO N
20040 A(J,K)=A(J,K)-D*A(I,K)
20050 NEXT K
20060 B(J)=B(J)-D*B(I)
20070 NEXT J,I
20080 X(N)=B(N)/A(N,N)
20090 FOR I=1 TO N-1
20100 J=N-I
```

```
20110 D=0
20120 FOR K=J+1 TO N
20130 D=D+A(J,K)*X(K)
20140 NEXT K
20150 X(J)=B(J)/A(J,J)-D/A(J,J)
20160 NEXT I
20170 RETURN
30000 END
```

```
FLOW IN PIPE NETWORKS,DARCY-WEISBACH VERSION

NO. OF ITERATIONS= 3

PIPE   LENGTH    DIAMETER  DISCHARGE  HEAD LOSS

  1   182.000      .300  +     .121  +    2.193
  2   182.000      .300  +     .050  +     .381
  3   182.000      .250  +     .037  +     .601
  4   182.000      .250  +     .023  +     .232
  5   182.000      .300  +     .021  +     .067
  6   182.000      .300  +     .004  +     .002
  7   122.000      .200  +     .058  +    3.450
  8   122.000      .250  +     .080  +    1.865
  9   122.000      .200  +     .035  +    1.251
 10   122.000      .200  +     .021  +     .450
 11   122.000      .250  +     .017  +     .084
 12   122.000      .200  +     .012  +     .151
```

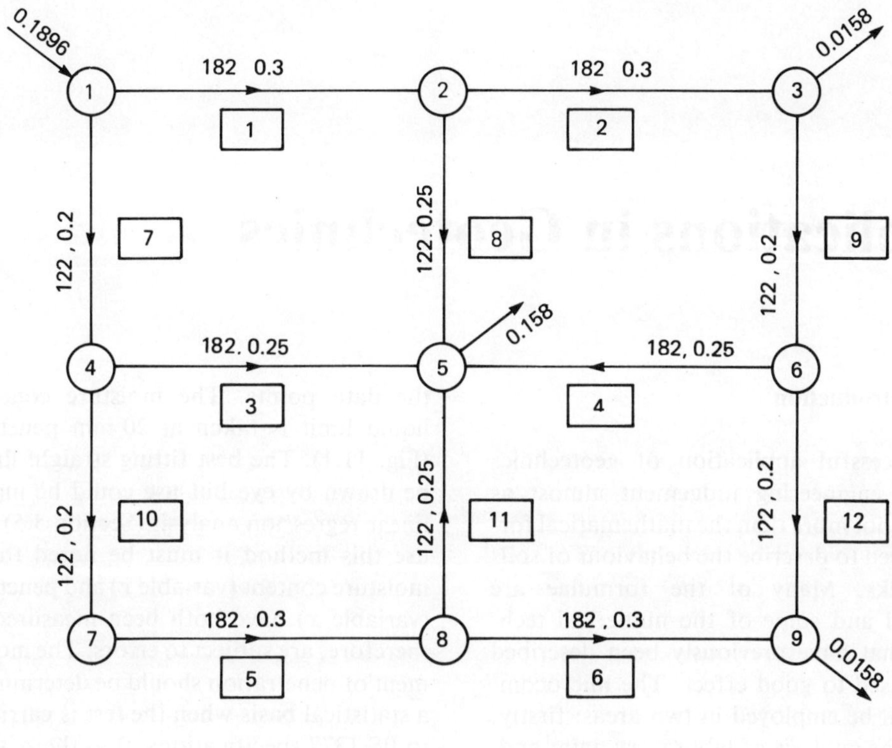

Fig. 10.4 Pipe network.

Example 10.2

The network shown in Fig. 10.4 consists of 12 pipes and 9 junctions numbered as indicated, and with the lengths and diameters as shown alongside each pipe. There is an inflow at junction 1 and outflows at junctions 3, 5, and 9. It is required to determine the flows in all pipes.

The data for this example are given in statements 10–24 and these consist of the following information:

Number of pipes
Number of junctions
For each pipe:
- Length (metres)
- Diameter (metres)
- Assumed discharge (m^3/s)
- Friction coefficient
- Joint or node numbers
- Nodal flow at each joint (m^3/s)

In determining the order in which the joint numbers are to be given for each pipe, the direction of flow is initially assumed as being from I to J. The joint numbers are then assigned accordingly, and the results will reveal whether or not the initial assumptions are correct.

The results for this example follow the listing of the program. A check on the continuity of flow at each junction and the summation of head loss around each loop confirms that the results are satisfactory.

Further Reading

Chow, V. T. *Open Channel Hydraulics*. McGraw-Hill, 1959.

Francis, J. R. D. *Fluid Mechanics for Engineering Students*. Edward Arnold, London, 1975.

11

Applications in Geotechnics

11.1 Introduction

The successful application of geotechnics involves engineering judgement almost as much, if not more than the mathematical formulae used to describe the behaviour of soils and rocks. Many of the formulae are empirical and some of the numerical techniques that have previously been described can be used to good effect. The microcomputer can be employed in two areas: firstly, that of the analysis of laboratory data; and, secondly, in the analysis and design of civil engineering foundations and structures.

11.2 Analysis of Laboratory Data

Data obtained during laboratory testing of soil are analysed to obtain soil parameters for use in soil classification, foundation design or the prediction of foundation behaviour. In most cases some form of curve fitting is applied to the data to obtain the required parameters and it is in this area that the microcomputer can be successfully applied. A simple illustration is the determination of the liquid limit of a soil. In the laboratory the behaviour of a soil at various moisture contents is examined using either the Cassagrande equipment or the cone penetrometer. The cone penetrometer measures the amount of penetration of a standard cone into soil samples having various moisture contents. The amount of penetration is then plotted against the moisture content of the soil samples and a line of best fit is drawn through the data points. The moisture content at liquid limit is taken at 20 mm penetration (Fig. 11.1). The best fitting straight line can be drawn by eye but use could be made of linear regression analysis (Section 5.5). If we use this method it must be noted that the moisture content (variable y) and penetration (variable x) have both been measured and, therefore, are subject to errors. The measurement of penetration should be determined on a statistical basis when the test is carried out to BS 1377 specifications, thus there is likely to be a greater error in the measurement of the moisture content. It is, therefore, normal to assume that penetration is the independent

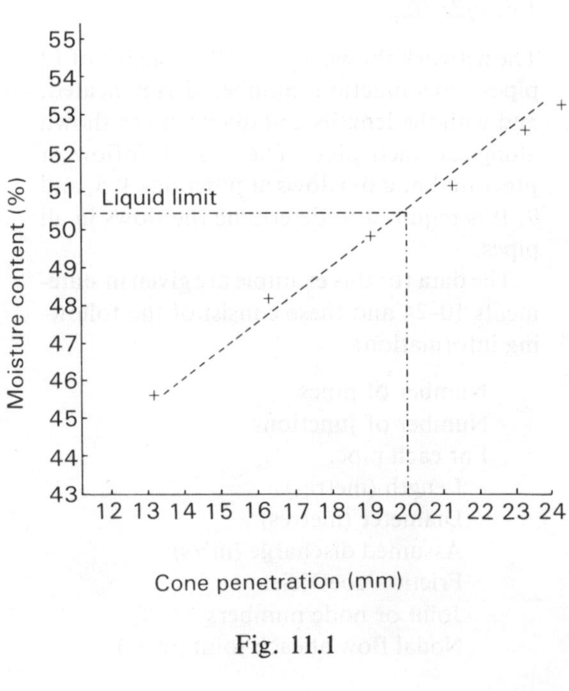

Fig. 11.1

variable and moisture content the dependent variable. The moisture content at liquid limit can then be obtained by calculating the regression of y on x.

The Cassagrande equipment enables a graph of moisture content versus the number of blows to be plotted. If both variables were plotted on arithmetic scales the resulting curve would be logarithmic. To obtain a linear relationship between our variables the number of blows is plotted on a logarithmic scale. A program which uses data from the Cassagrande test must therefore convert the x variable (number of blows) to logarithmic values prior to any regression analysis.

Regression analysis has the advantage of enabling statistical information in the form of the correlation coefficient and the stan-

Program 11.1 Listing and output

```
10 PRINT"**********************************"
20 PRINT   LIQUID LIMIT ANALYSIS"
30 REM
40 REM THIS PROGRAM WILL RUN ON EITHER
50 REM AN APPLE OR A COMMODORE  MICRO-
60 REM COMPUTER.
70 REM
80 PRINT"**********************************"
90 REM ****** DISPLAY INSTRUCTIONS ***
100 PRINT
110 PRINT" THIS PROGRAM DETERMINES THE LIQUID"
120 PRINT"LIMIT OF A SOIL USING THE RESULTS"
130 PRINT"FROM THE CASAGRANDE APPARATUS OR THE"
140 PRINT"CONE PENETRATION EQUIPMENT"
150 PRINT
160 PRINT"DATA"
170 PRINT"----"
180 PRINT"NUMBER OF BLOWS OR THE PENETRATION"
190 PRINT"AND THE MOISTURE CONTENT VALUES "
200 REM ****** SELECT EQUIPMENT ********
210 PRINT
220 PRINT "SELECT TYPE OF TEST"
230 PRINT "CASAGRANDE EQUIPMENT  : ENTER 1"
240 PRINT "CONE PENETROMETER     : ENTER 2"
250 INPUT TP
260 IF TP=1 THEN V=LOG(25)/LOG(10)
270 IF TP=2 THEN V=20
280 PRINT
290 PRINT "NUMBER OF KNOWN POINTS"
300 INPUT N
310 REM ****** INITIALIZE *************
320 J=0
330 K=0
340 L=0
350 M=0
360 R2=0
370 REM ****** INPUT DATA **************
```

```
380 FOR I=1 TO N
390 PRINT"POINT";I
400 IF TP=2 THEN GOTO 440
410 PRINT"NUMBER OF BLOWS"
420 INPUT X
430 IF TP=1 THEN GOTO 460
440 PRINT"PENETRATION (MM)"
450 INPUT X
460 PRINT"MOISTURE CONTENT"
470 INPUT Y
480 IF TP=1 THEN X=LOG(X)/LOG(10)
490 J=J+X
500 K=K+Y
510 L=L+X↑2
520 M=M+Y↑2
530 R2=R2+X*Y
540 NEXT I
550 B=(N*R2-K*J)/(N*L-J↑2)
560 A=(K-B*J)/N
570 REM ***** OUTPUT RESULTS *********
580 PRINT
590 PRINT
600 PRINT "RESULTS"
610 PRINT "-------"
620 X1=V
630 Y=A+B*X1
640 PRINT"LIQUID LIMIT = ";Y
650 PRINT
660 PRINT "F(X) =";A;"+ (";B;"* X)"
670 J=B*(R2-J*K/N)
680 M=M-K↑2/N
690 K=M-J
700 PRINT
710 R2=J/M
720 PRINT "COEFFICIENT OF ";
730 PRINT "DETERMINATION";
740 PRINT "(R↑2)= ";R2
750 PRINT
760 PRINT "COEFFICIENT OF ";
770 PRINT "CORRELATION=";SQR(R2)
780 PRINT
790 PRINT"STANDARD ERROR OF EST";
800 PRINT "IMATE=";SQR(K/(N-2))
810 PRINT
820 PRINT
830 PRINT"DO YOU REQUIRE TO CALCULATE"
840 PRINT"ANOTHER RESULT( YES OR NO)"
850 INPUT A$
860 IF LEFT$(A$,1)="Y" THEN 220
870 END
```

```
*********************************
   LIQUID LIMIT ANALYSIS
*********************************

 THIS PROGRAM DETERMINES THE LIQUID
LIMIT OF A SOIL USING THE RESULTS
FROM THE CASAGRANDE APPARATUS OR THE
CONE PENETRATION EQUIPMENT

DATA
----
NUMBER OF BLOWS OR THE PENETRATION
AND THE MOISTURE CONTENT VALUES

SELECT TYPE OF TEST
CASAGRANDE EQUIPMENT : ENTER 1
CONE PENETROMETER    : ENTER 2
 1

NUMBER OF KNOWN POINTS
 4
POINT 1
NUMBER OF BLOWS
 48
MOISTURE CONTENT
 81.4
POINT 2
NUMBER OF BLOWS
 29
MOISTURE CONTENT
 83.7
POINT 3
NUMBER OF BLOWS
 20
MOISTURE CONTENT
 85.5
POINT 4
NUMBER OF BLOWS
 10
MOISTURE CONTENT
 88.9

RESULTS
-------
LIQUID LIMIT = 84.4692781

F(X) = 99.8930803 + (-11.0332361 * X)
```

```
COEFFICIENT OF DETERMINATION(R↑2)=   .999678925

COEFFICIENT OF CORRELATION= .99983945

STANDARD ERROR OF ESTIMATE= .0694532343

DO YOU REQUIRE TO CALCULATE
ANOTHER RESULT( YES OR NO)

NO

READY.
```

dard error of estimate to be used to assess the validity of the data.

Program 11.1 shows a listing of a program which calculates the liquid limit of a clay soil using data from either the cone penetrometer or the Cassagrande equipment. The program outputs the correlation coefficient and the standard error of estimate as well as the liquid limit as shown in the trial run. Three or more sets of data points are necessary to obtain meaningful results.

Linear regression analysis can also be used to determine the soil shear strength parameters from triaxial testing. The normal way to determine these parameters is by drawing Mohr circles for each set of triaxial test results and then constructing the failure envelope which is approximately tangential to them all. There are a variety of methods that calculate the Mohr failure envelope which is assumed to be linear (Franklin, 1971; Hubert and Rubey, 1959). A relatively simple method exists using the concept of stress paths rather than Mohr circles to calculate the linear failure envelope. The American method of stress representation which can be used to plot stress paths represents stress on a sample as a single point on a diagram whose coordinates are:

$$p = \frac{\sigma_1 + \sigma_3}{2} \qquad (11.1)$$

and

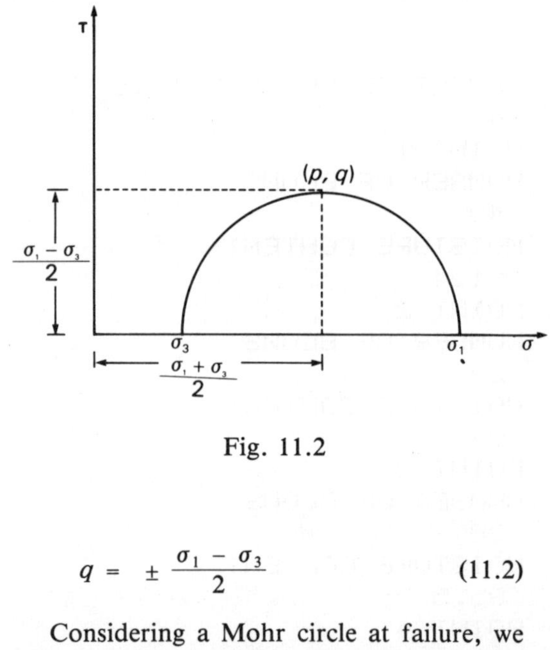

Fig. 11.2

$$q = \pm \frac{\sigma_1 - \sigma_3}{2} \qquad (11.2)$$

Considering a Mohr circle at failure, we can represent the σ_1 and σ_3 stresses at failure by one point. As can be seen from Fig. 11.2, this point is the uppermost point on the circle. Point p is the x coordinate for the centre of the Mohr circle and q is one-half of the deviator stress. Hence a q–p diagram can be drawn (rather than a Mohr circle diagram) with the stresses at failure represented by points. A modified failure envelope can then be drawn through these points by linear regression analysis (Fig. 11.3).

The intercept and slope of this line are

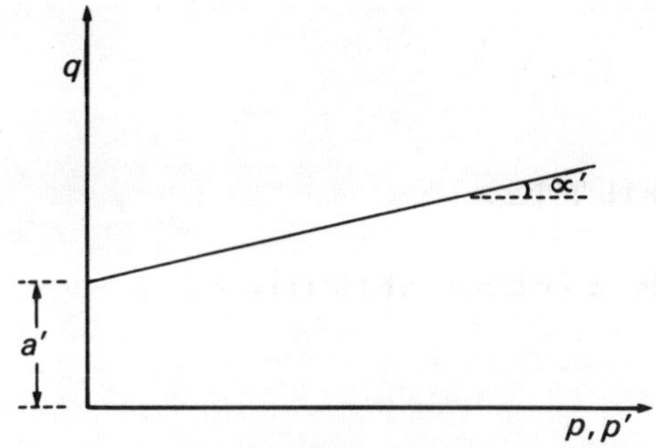

Fig. 11.3

related to the actual angle of friction ϕ' and cohesion c' by the following equations:

$$\phi' = \sin^{-1}(\tan\alpha') \tag{11.3}$$

$$c' = \frac{a'}{\cos\phi'} \tag{11.4}$$

A program to calculate friction and cohesion by this method is provided in Program 11.2. Only the cell pressure and the deviator stress for a minimum of three triaxial tests is required. The program can be used to calculate the undrained shear strength parameter

Program 11.2 Listing and output

```
10 PRINT"********************************"
20 PRINT   SOIL STRENGTH PARAMETERS
30 REM
40 REM THIS PROGRAM WILL RUN ON EITHER
50 REM AN APPLE OR A COMMODORE MICRO-
60 REM COMPUTER.
70 REM
80 PRINT"********************************"
90 REM ****** PRINT INSTRUCTIONS *****
100 PRINT
110 PRINT"THIS PROGRAM DETERMINES THE ANGLE "
120 PRINT"OF FRICTION AND COHESION OF A SOIL"
130 PRINT"FROM THE TRIAXIAL DATA."
140 PRINT"ONLY THE CELL PRESSURE AND THE MAX.  "
150 PRINT"DEVIATOR STRESS FOR EACH TEST"
160 PRINT"IS REQUIRED"
170 PRINT
180 PRINT"NOTE"
190 PRINT"IF EFFECTIVE STRESS VALUES ARE "
200 PRINT"REQUIRED THEN THE CELL PRESSURE"
210 PRINT"MUST BE REDUCED BY THE PORE WATER"
220 PRINT"PRESSURE AT FAILURE. THE DEVIATOR "
230 PRINT"STRESS WILL REMAIN THE SAME"
240 REM ****** INPUT DATA **************
250 PRINT
260 PRINT"ENTER NO OF TRIAXIAL TESTS"
270 INPUT N
280 J=0
290 K=0
300 L=0
310 M=0
320 R2=0
```

```
330 PRINT
340 FOR I=1 TO N
350 PRINT
360 PRINT"TRIAXIAL DATA TEST";I
370 PRINT"ENTER CELL PRESSURE (KN/M↑2) "
380 INPUT CP
390 PRINT"ENTER MAXIMUM DEVIATOR STRESS (KN/M↑2)"
400 INPUT DS
410 Y=DS/2
420 X=CP+Y
430 J=J+X
440 K=K+Y
450 L=L+X↑2
460 M=M+Y↑2
470 R2=R2+X*Y
480 NEXT I
490 REM ****** REGRESSION ANALYSIS ****
500 B=(N*R2-K*J)/(N*L-J↑2)
510 A=(K-B*J)/N
520 REM ****** OUTPUT RESULTS *********
530 PRINT
540 PRINT"RESULTS"
550 PRINT"-------"
560 PRINT"STRESS PATH ENVELOPE(AMERICAN METHOD"
570 B1=INT(ATN(B)+.5)*(180/π)
580 PRINT"ANGLE ALPHA = ";B1
590 PRINT"INTERCEPT A = ";A
600 J=B*(R2-J*K/N)
610 M=M-K↑2/N
620 K=M-J
630 PRINT
640 R2=J/M
650 PRINT "COEFFICIENT OF ";
660 PRINT "DETERMINATION";
670 PRINT "(R↑2) = ";R2
680 PRINT
690 PRINT "COEFFICIENT OF ";
700 PRINT "CORRELATION =";SQR(R2)
710 PRINT
720 PRINT"STANDARD ERROR OF EST";
730 PRINT "IMATE =";SQR(K/(N-2))
740 PRINT
750 X1=V
760 Y=A+B*X1
770 PH=INT(ATN(B/SQR(-B*B+1))*(180/π)+.5)
780 CO=A/(COS(PH*π/180))
790 PRINT
800 PRINT"SOIL SHEAR STRENGTH PARAMETERS"
810 PRINT"COHESION = ";CO
```

```
820 PRINT"ANGLE OF SOIL FRICTION = ";PH
830 PRINT
840 PRINT
850 PRINT"DO YOU REQUIRE TO CALCULATE"
860 PRINT"ANOTHER RESULT(YES OR NO)"
870 INPUT A$
880 IF LEFT$(A$,1)="Y" THEN250
890 END
```

```
*********************************
     SOIL STRENGTH PARAMETERS
*********************************

THIS PROGRAM DETERMINES THE ANGLE
OF FRICTION AND COHESION OF A SOIL
FROM THE TRIAXIAL DATA.
ONLY THE CELL PRESSURE AND THE MAX.
DEVIATOR STRESS FOR EACH TEST
IS REQUIRED

NOTE
IF EFFECTIVE STRESS VALUES ARE
REQUIRED THEN THE CELL PRESSURE
MUST BE REDUCED BY THE PORE WATER
PRESSURE AT FAILURE. THE DEVIATOR
STRESS WILL REMAIN THE SAME

ENTER NO OF TRIAXIAL TESTS
 3

TRIAXIAL DATA TEST 1
ENTER CELL PRESSURE (KN/M↑2)
 140
ENTER MAXIMUM DEVIATOR STRESS (KN/M↑2)
 238.2

TRIAXIAL DATA TEST 2
ENTER CELL PRESSURE (KN/M↑2)
 280
ENTER MAXIMUM DEVIATOR STRESS (KN/M↑2)
 250.7

TRIAXIAL DATA TEST 3
ENTER CELL PRESSURE (KN/M↑2)
 420
ENTER MAXIMUM DEVIATOR STRESS (KN/M↑2)
 265.4
```

```
RESULTS
-------

STRESS PATH ENVELOPE(AMERICAN METHOD
ANGLE ALPHA =  0
INTERCEPT A =  106.921441

COEFFICIENT OF DETERMINATION(R↑2) =  .998021267

COEFFICIENT OF CORRELATION = .999010144

STANDARD ERROR OF ESTIMATE = .428243221


SOIL SHEAR STRENGTH PARAMETERS
COHESION =  107.068175
ANGLE OF SOIL FRICTION =  3


DO YOU REQUIRE TO CALCULATE
ANOTHER RESULT(YES OR NO)
NO
```

where ϕ' is close to zero. The same program can be used to obtain effective shear strength parameters by subtracting the pore water pressure at failure from the cell pressure, but the deviator stress will not require alteration.

One problem with the method outlined above is the difficulty of obtaining true confidence limits on the slope tan ϕ' and the intercept c'. This can be overcome if the technique described by Lisle and Strom (1982) is used. Their method fits a linear Mohr envelope to Mohr circles by minimizing the sum of the squares of the radial distances from the Mohr circles perpendicular to the best-fit linear Mohr envelope. As the linear Mohr envelope is obtained directly it is much easier to obtain the required confidence limits.

11.3 Design

The economic design of foundations and earth structures often involves consideration of many variables. In particular there will be some doubt as to the accuracy and reliability of some of the soil parameters themselves. A standard 'limit state' approach to design (as with CP 110) would help to overcome this doubt by using probability theory. However, it will be several years before this is a standard method in geotechnics. Microcomputers can, therefore, be used in parametric studies to investigate the influence of specific parameters on a design. A simple example to illustrate this would be to examine the effect of soil friction and wall friction on the depth of penetration and height of soil retained by a cantilevered sheet pile retaining wall.

11.3.1 Cantilevered Sheet Pile Wall

Let us assume that we have a cantilevered sheet pile wall retaining a cohensionless soil to a height H with the pile penetrating this soil to a depth l with the point of rotation when failure occurs being at depth d. The wall and the assumed design pressure distributions are shown in Fig. 11.4.

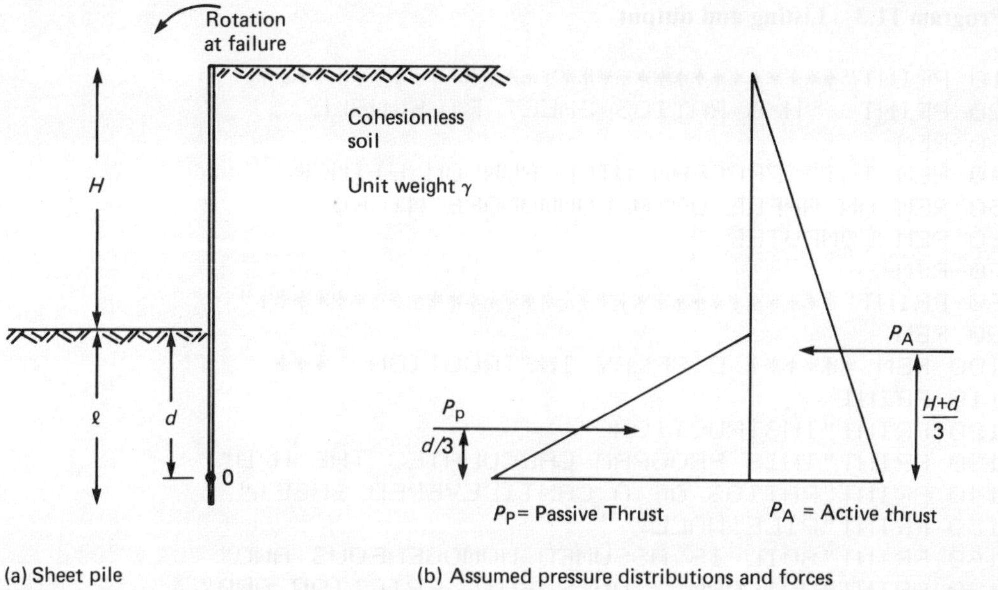

Fig. 11.4

The stability of a cantilevered sheet pile wall is due entirely to the passive resistance of the soil in front of the wall. Failure occurs if the passive resistance is overcome and the wall will rotate about a point near its base (point 0). If we consider the pile below this point, passive resistance will develop beneath this level behind the pile when failure occurs. The design assumptions ignore this passive resistance and the distance d is determined by considering the moments about 0. The length l is obtained by increasing d by 20% to allow for the development of passive resistance behind the wall.

The total active and passive thrusts on the sheet pile wall are given by:

$$P_A = \tfrac{1}{2}\gamma K_A (H + d)^2 \qquad (11.5)$$

and

$$P_P = \tfrac{1}{2}\gamma K_P d^2 \qquad (11.6)$$

where K_A is the active earth pressure coefficient and K_P is the passive earth pressure coefficient.

The values of K_A and K_P are affected by wall friction (δ) and the following formula can be used to determine K_A and K_P for a vertical wall with horizontal ground surface behind the wall.

$$K_A = \frac{\sin^2(90 + \phi)}{\sin(90 - \delta)\left[1 + \sqrt{\left(\dfrac{\sin(\phi + \delta)\sin\phi}{\sin(90 - \delta)}\right)}\right]^2} \qquad (11.7)$$

and

$$K_P = \frac{\sin^2(90 - \phi)}{\sin(90 + \delta)\left[1 - \sqrt{\left(\dfrac{\sin(\phi + \delta)\sin\phi}{\sin(90 + \delta)}\right)}\right]^2} \qquad (11.8)$$

Taking moments about point 0 we obtain:

$$\tfrac{1}{2}\gamma K_A (H + d)^2 (H + d)/3 = \tfrac{1}{2}\gamma K_P d^2 d/3 \qquad (11.9)$$

which simplifies to:

$$(H + d)^3 = (K_P/K_A)d^3 \qquad (11.10)$$

$$\therefore H + d = d(K_P/K_A)^{\frac{1}{3}} \qquad (11.11)$$

Thus

$$H/d = (K_P/K_A)^{\frac{1}{3}} - 1 \qquad (11.12)$$

The ratio H/d can thus be determined for

Program 11.3 Listing and output

```
10 PRINT"********************************"
20 PRINT   H/D RATIOS SHEET PILE WALL
30 REM
40 REM THIS PROGRAM WILL RUN ON EITHER
50 REM AN APPLE OR A COMMODORE MICRO-
60 REM COMPUTER.
70 REM
80 PRINT"********************************"
90 REM
100 REM ****** DISPLAY INSTRUCTIONS ***
110 PRINT
120 PRINT"INSTRUCTIONS"
130 PRINT"THIS PROGRAM CALCULATES THE H/D"
140 PRINT"RATIOS OF A CANTILEVERED SHEET"
150 PRINT"PILE WALL"
160 PRINT"SOIL IS ASSUMED HOMOGENEOUS AND"
170 PRINT"UNIFORM.  ONLY SOIL FRICTION AND"
180 PRINT"WALL FRICTION IS REQUIRED AS DATA"
190 PRINT
200 PRINT"ENTER DATA"
210 PRINT"SOIL FRICTION"
220 INPUT PH
230 PRINT"WALL FRICTION"
240 INPUT D
250 REM *** CALCULATE ACTIVE AND PASSIVE ***
260 REM *** EARTH PRESSURE COEFFICIENTS  ***
270 G=1.74533E-2
280 K1=(SIN((90+PH)*G))↑2
290 K2=1+SQR((SIN((PH+D)*G)*SIN(PH*G))/SIN((90-D)*G))
300 K3=SIN((90-D)*G)*K2↑2
310 KA=K1/K3
320 K1=(SIN((90-PH)*G))↑2
330 K2=1-SQR((SIN((PH+D)*G)*SIN(PH*G))/SIN((90+D)*G))
340 K3=SIN((90+D)*G)*K2↑2
350 KP=K1/K3
360 REM ****** CALCULATE H/D RATIO ***
370 HD=((KP/KA)↑.33333)-1
380 REM ****** OUTPUT ****************
390 PRINT
400 PRINT"      RESULTS"
410 PRINT"      -------"
420 PRINT"SOIL FRICTION =";PH
430 PRINT"WALL FRICTION =";D
440 PRINT"ACTIVE COEFF. ";KA
450 PRINT"PASSIVE COEFF. ";KP
460 PRINT"H/D RATIO ";HD
470 PRINT
```

```
480 PRINT"DO YOU REQUIRE ANOTHER RUN"
490 PRINT"ANSWER YES OR NO"
500 INPUT A$
510 IF LEFT$(A$,1)="Y" THEN 200
520 REM ******** END *****************
530 END
```

```
*******************************
  H/D RATIOS SHEET PILE WALL
*******************************

INSTRUCTIONS
THIS PROGRAM CALCULATES THE H/D
RATIOS OF A CANTILEVERED SHEET
PILE WALL
SOIL IS ASSUMED HOMOGENEOUS AND
UNIFORM.  ONLY SOIL FRICTION AND
WALL FRICTION IS REQUIRED AS DATA

ENTER DATA
SOIL FRICTION
 30
WALL FRICTION
 15

    RESULTS
    -------
SOIL FRICTION = 30
WALL FRICTION = 15
ACTIVE COEFF.    .301416221
PASSIVE COEFF.  4.976511
H/D RATIO  1.54633412

DO YOU REQUIRE ANOTHER RUN
ANSWER YES OR NO
YES
ENTER DATA
SOIL FRICTION
 30
WALL FRICTION
 10

    RESULTS
    -------
SOIL FRICTION = 30
WALL FRICTION = 10
```

```
ACTIVE COEFF.     .308465362
PASSIVE COEFF.   4.14330706
H/D RATIO   1.37707643

DO YOU REQUIRE ANOTHER RUN
ANSWER YES OR NO
NO
```

various values of soil and wall friction (Program 11.3). These data can then be used to produce a design chart as shown in Fig. 11.5. Although this is a simple example, Fig. 11.5 shows how wall friction affects the penetration depth of a cantilevered sheet pile wall. It is important to note that the data in Fig. 11.5 is for a factor of safety of 1.0. When a factor of safety is introduced it is applied to the coefficient of passive earth pressure, i.e.

$$H/d = ((K_P/F)/K_A)^{\frac{1}{3}} - 1 \qquad (11.13)$$

The design chart in Fig. 11.6 is for a safety factor of 1.5. Note the differences in values of H/d for a particular ϕ and δ.

11.3.2 Slope Stability

The determination of the factor of safety for a slope is a tedious process and, therefore, eminently suitable for computer application. There are essentially three main methods of slope stability using limiting equilibrium techniques. These are:

(a) Fellenius's method or 'ordinary method of slices';
(b) Bishop's simplified method;
(c) Morgenstern and Price's method.

Application of all these methods to the same slopes (stable and failed) indicates that

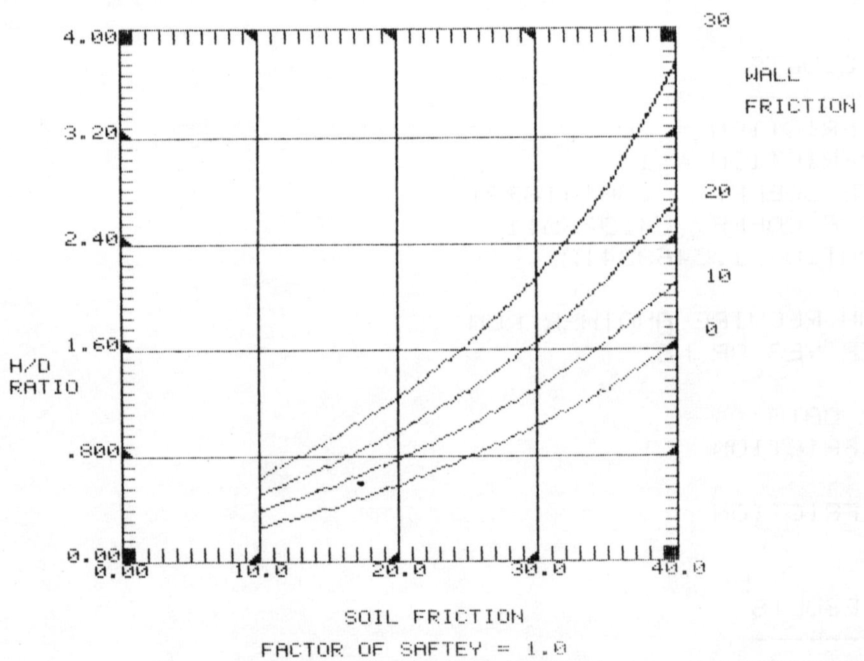

Fig. 11.5 Sheet pile H/D ratios.

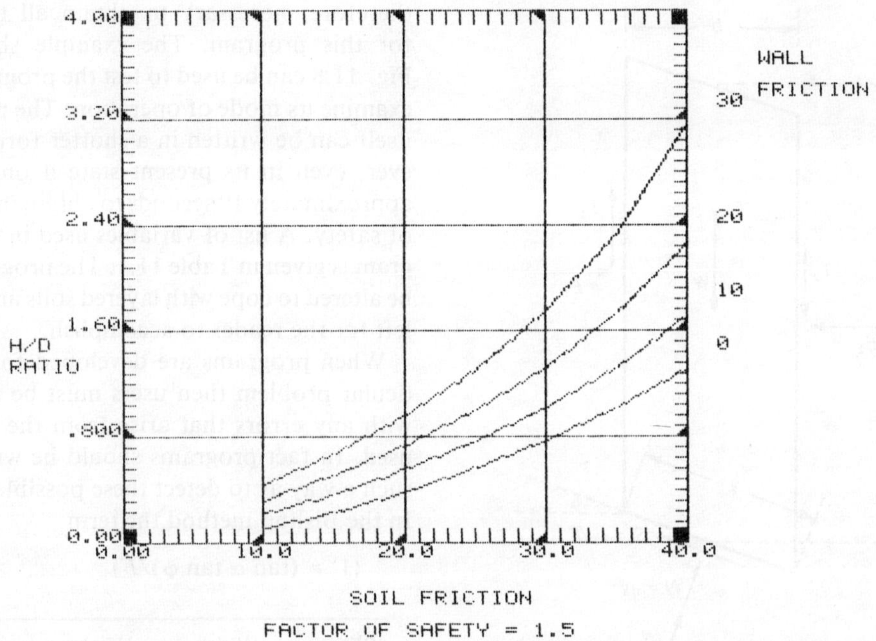

Fig. 11.6 Sheet pile H/D ratios.

the Morgenstern and Price method is possibly the most accurate, however, it is also the most difficult to use. Evidence suggests that the Fellenius method underestimates the factor of safety and is, therefore, conservative. The extent of underestimation can in certain circumstances produce uneconomic slope design. The Morgenstern and Price method can give a range of values for the safety factor (depending on the assumptions made), with the Bishop simplified solution producing safety factor values within or close to the range specified. It should be noted that different methods will in general lead to different positions of the critical slip surface.

The choice of method and identification of a possible slip surface is dependent on the experience of the engineer and there are various factors that one should be conversant with which need careful attention.

Bishop's Simplified Method. Bishop (1955) formulated the problem rigorously by including the inter-slice forces in the equations of equilibrium of a single slice. He demonstrated that a simplified calculation can be made giving fairly accurate results even though certain inter-slice forces are ignored. The normal stress on the failure plane for a single slice is obtained by resolving the forces vertically instead of normal to the failure surface as in the Fellenius method (Fig. 11.7).

The factor of safety for a slope is given by

$$F = \frac{\Sigma\,(c'b + (W - ub)\tan\phi')/m_\alpha}{\Sigma\,W\sin\alpha} \qquad (11.14)$$

where $m_\alpha = (1 + (\tan\alpha\tan\phi)/F)\cos\alpha$, c' = cohesion, and ϕ' = soil friction.

Program 11.4 calculates the factor of safety for a slope in homogeneous soil by the Bishop method but uses the pore water pressure ratio instead of the pore water pressure on each slice.

The program requests details of the soil in the slope, the number of slices and their dimensions and an approximation of the factor of safety which is a requirement of the Bishop method. A scale drawing showing the section of the slope and the slip circle is,

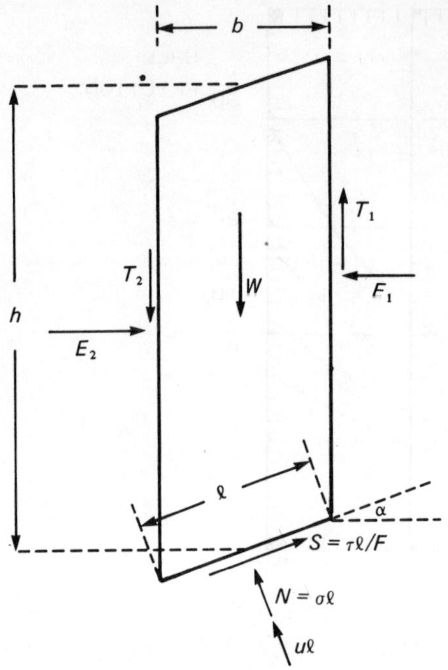

Fig. 11.7 b = width of slice, h = height of slice, l = length of failure surface, α = angle of failure surface to horizontal, W = weight of slice S = shear force along base of slice, N = normal force on base of slice, ul = pore water force, T_1 and T_2 = shear forces on sides of slice, E_1 and E_2 = normal forces on sides of slice, F = factor of safety.

therefore, necessary to obtain all the data for this program. The example shown in Fig. 11.8 can be used to test the program and examine its mode of operation. The program itself can be written in a shorter form, however, even in its present state it only takes approximately 10 seconds to obtain the factor of safety. A list of variables used in the program is given in Table 11.1. The program can be altered to copc with layered soils and this is left for the reader to accomplish.

When programs are developed for a particular problem then users must be familiar with any errors that arise from the method used. In fact programs should be written in such a way as to detect these possible errors. In the Bishop method the term

$$(1 + (\tan \alpha \tan \phi)/F)$$

Slice No.	Width (m)	Height (m)	Angle (°)
1	2.4	1.9	57
2	2.0	4.4	38.5
3	2.0	5.0	23.8
4	2.0	4.6	11
5	2.0	3.8	0.5
6	2.0	2.5	–12
7	1.9	0.9	–24

(Fig. 11.8)

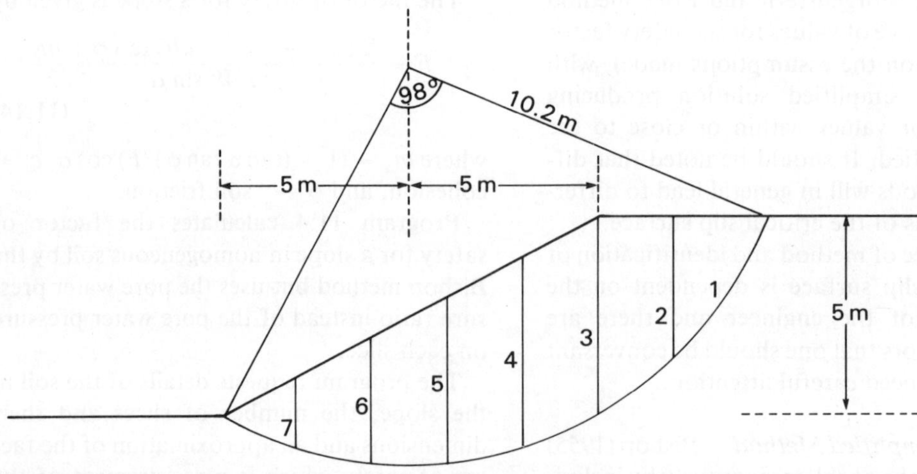

Fig. 11.8 Soil properties: γ = 19 kN/m³, c' = 5 kN/m², ϕ' = 20°, r_u = 0. Factor of safety = 1.662.

Program 11.4 Listing and output

```
10 REM ************************************
20 REM     SLOPE STABILITY PROGRAM
30 REM
40 REM  THIS PROGRAM WILL RUN ON EITHER
50 REM  AN APPLE OR A COMMODORE MICRO-
60 REM  COMPUTER
70 REM
80 REM ************************************
90 DIM F(8,7)
100 REM *** G1 IS A CONVERSION FACTOR:DEG TO RADS ***
110 G1=1.74533E-02
120 A1=0
130 P1=0
140 PRINT"SLOPE STABILITY ANALYSIS"
150 PRINT"========================"
160 PRINT
170 PRINT"THIS PROGRAM DETERMINES THE FACTOR OF"
180 PRINT"SAFETY FOR A SLOPE IN HOMOGENEOUS SOIL"
190 PRINT"USING THE BISHOP SIMPLIFIED METHOD."
200 PRINT"PORE WATER PRESSURES ARE EXPRESSED AS"
210 PRINT"THE PORE PRESSURE RATIO.
220 PRINT"SOIL DETAILS AND SLICE INFORMATION IS"
230 PRINT"ALSO REQUIRED"
240 REM ****** INPUT SOIL DATA *********
250 PRINT
260 PRINT
270 PRINT"ENTER THE FOLLOWING DATA"
280 PRINT
290 PRINT"SOIL UNIT WEIGHT(KN/M↑3)"
300 INPUT UW
310 PRINT"COHESION (KN/M↑2)"
320 INPUT CO
330 PRINT"SOIL FRICTION (DEGREES)"
340 INPUT PI
350 PRINT"PORE WATER PRESS. RATIO"
360 INPUT RU
370 PRINT
380 PRINT"NUMBER OF SLICES"
390 INPUT NS
400 IF NS<30 THEN GOTO 440
410 PRINT
420 PRINT"TOO MANY SLICES ONLY 30 ALLOWED"
430 GOTO 380
440 REM  ****** INPUT SLICE DATA ********
450 PRINT
460 DIM B(NS),H(NS),A(NS),W(NS),WK(NS),PW(NS),CB(NS),MA(NS)
470 FOR I=1 TO NS
```

```
480 REM ***CALL SUBROUTINE FOR SLICE INPUT ***
490 GOSUB 840
500 NEXT I
510 REM *** CHECK SLICE DATA ***
520 GOSUB 940
530 PRINT
540 PRINT"ENTER INITIAL FACTOR OF SAFETY"
550 INPUT F1
560 PRINT
570 REM ****** CALCULATE NEW SAFETY FACTOR ******
580 PRINT"COMPUTING"
590 S1=0
600 S2=0
610 FOR I=1 TO NS
620 WT=B(I)*H(I)*UW
630 W(I)=WT*(SIN(A(I)*G1))
640 S1=S1+W(I)
650 PW(I)=RU*UW*H(I)*B(I)
660 WK(I)=((WT-PW(I))*TAN(PI*G1))+(CO*B(I))
670 MA(I)=1+((TAN(PI*G1)*TAN(A(I)*G1))/F1)
680 TP=1/(COS(A(I)*G1))
690 MA(I)=TP/MA(I)
700 TP=MA(I)*WK(I)
710 S2=S2+TP
720 NEXT I
730 F2=S2/S1
740 REM ****** CHECK ESTIMATED WITH CALCULATED ******
750 IF ABS(F1-F2)<.005 THEN GOTO 63999
760 REM *** SET ESTIMATED F TO NEW CALCULATED VALUE"
770 F1=F2
780 GOTO 590
790 REM ****** OUTPUT SAFETY FACTOR ******
800 PRINT
810 PRINT"FACTOR OF SAFETY OF SLOPE = ";F2
820 END
830 REM *** SLICE DATA INPUT SUBROUTINE ***
840 PRINT
850 PRINT"SLICE NUMBER";I
860 PRINT"WIDTH (M)   "
870 INPUT B(I)
880 PRINT"HEIGHT (M)   "
890 INPUT H(I)
900 PRINT"ANGLE OF BASE"
910 INPUT A(I)
920 PRINT
930 RETURN
940 REM **** SUBROUTINE TO CHECK SLICE DATA ***
950 PRINT
960 PRINT"SLICE DATA"
```

```
970 PRINT"=========="
980 PRINT"SLICE N0     WIDTH    HEIGHT     ANGLE"
990 PRINT"--------     -----    ------     -----"
1000 PRINT
1010 FOR I=1 TO NS
1020 PRINT I,B(I),H(I),A(I)
1030 NEXT I
1040 PRINT
1050 PRINT"DO YOU REQUIRE TO ALTER ANY SLICE DATA "
1060 PRINT"ENTER YES OR NO"
1070 PRINT
1080 INPUT AN$
1090 IF LEFT$(AN$,1)<>"Y" THEN RETURN
1100 PRINT"PLEASE ENTER SLICE NUMBER"
1110 INPUT I
1120 IF I<NS THEN GOTO 1170
1130 PRINT
1140 PRINT"SLICE NUMBER MUST BE LESS THAN ";NS
1150 PRINT
1160 GOTO 1100
1170 GOSUB 840
1180 GOTO 960
```

```
SLOPE STABILITY ANALYSIS
========================

THIS PROGRAM DETERMINES THE FACTOR OF
SAFETY FOR A SLOPE IN HOMOGENEOUS SOIL
USING THE BISHOP SIMPLIFIED METHOD.
PORE WATER PRESSURES ARE EXPRESSED AS
THE PORE PRESSURE RATIO.
SOIL DETAILS AND SLICE INFORMATION IS
ALSO REQUIRED


ENTER THE FOLLOWING DATA

SOIL UNIT WEIGHT(KN/M↑3)
 19
COHESION (KN/M↑2)
 5
SOIL FRICTION (DEGREES)
 20
PORE WATER PRESS. RATIO
 0
NUMBER OF SLICES
 7
```

```
SLICE NUMBER 1
WIDTH (M)
 2.4
HEIGHT (M)
 1.9
ANGLE OF BASE
 57

SLICE NUMBER 2
WIDTH (M)
 2
HEIGHT (M)
 4.4
ANGLE OF BASE
 38.5

SLICE NUMBER 3
WIDTH (M)
 2
HEIGHT (M)
 5
ANGLE OF BASE
 23.8

SLICE NUMBER 4
WIDTH (M)
 2
HEIGHT (M)
 4.6


ANGLE OF BASE
 11

SLICE NUMBER 5
WIDTH (M)
 2
HEIGHT (M)
 3.8
ANGLE OF BASE
 .5
```

```
SLICE NUMBER 6
WIDTH (M)
 2
HEIGHT (M)
 2.5
ANGLE OF BASE
-12

SLICE NUMBER 7
WIDTH (M)
 1.9
HEIGHT (M)
 .9
ANGLE OF BASE
-24

SLICE DATA
==========
SLICE NO     WIDTH    HEIGHT        ANGLE
========     =====    ======        =====

 1            2.4          1.9            57
 2            2            4.4          38.5
 3            2            5            23.8
 4            2            4.6            11
 5            2            3.8            .5
 6            2            2.5           -12
 7            1.9           .9           -24

DO YOU REQUIRE TO ALTER ANY SLICE DATA
ENTER YES OR NO

NO

ENTER INITIAL FACTOR OF SAFETY
 3

COMPUTING

FACTOR OF SAFETY OF SLOPE =  1.66186541

READY.
```

(which is used to obtain m_α) can become negative near the toe of a steeply inclined slip surface. This can happen because α is often a negative angle, thus, if

$$\alpha = \tan^{-1}(\tan\phi/F) - 90$$

then

$$m_\alpha \rightarrow 0$$

As the term approaches zero the safety factor is overestimated because the normal force at the base of the slice approaches infinity. If the slip surface becomes steeper then the value of F is underestimated because negative values of normal force on the base of a slice introduce an incorrect direction for shear resistance. Programs should, therefore, check on the value of m_α and when it becomes smaller than 0.2 a message should be printed out warning that the value of F should be used with caution.

Morgenstern and Price Method. This method is probably the most general approach for limit equilibrium analysis of a slope. The inter-slice forces (T_1 and T_2, E_1 and E_2) are taken into account. In order to make the problem statically determinate a mathematical relationship between the forces E_1 and E_2 and T_1 and T_2 is assumed. Hand calculations are, therefore, impracticable and program complexity puts it beyond the scope of this text.

Table 11.1 Slope Stability Program: Listing of Variables Used

UW	Soil unit weight
CO	Cohesion
PI	Soil friction
RU	Pore water pressure ratio
NS	Number of slices
B(*I*)	Width of slices array
H(*I*)	Height of slices array
A(*I*)	Angle of base of slice array
FT	Initial factor of safety
WT	Weight of a slice
WK(*I*)	Work space array
PW(*I*)	Pore water pressure array
CB(*I*)	Work space array (cohesion *slice width)
MA(*I*)	m_α
TP	Temporary working value
ST	Sum variable
S2	Sum variable
F2	Calculated factor of safety

Back Analysis. Program 11.4 can also be used to analyse a failed slope. When a slope has failed, the failure surface is known and the program can be used to estimate the average shear strength parameters of the slope. This can be accomplished by entering various combinations of soil friction and cohesion to obtain the factor of safety of 1.0. The example shown in Fig. 11.9 can be used for this purpose. Note it is unnecessary to enter slice data more than once. If the program is run with the first soil friction and cohesion values, on completion the computer will exit from the program. The variables for soil friction (*PI*) and cohesion (*CO*) can then be altered directly from the keyboard and the program rerun by a GOTO 54Ø. Alternatively the program can be modified to ask for alterations at the end of a run prior to any repeat calculations.

11.4 Rate of Consolidation Settlement

The rate of consolidation settlement was first analysed by Terzaghi who produced a theory which derived the following differential equation of consolidation.

$$C_v \frac{\partial^2 u}{\partial z^2} = \frac{\partial u}{\partial t} \tag{11.15}$$

where $C_v = k/\gamma_w m_v$ (coefficient of consolidation), u = pore water pressure, z = depth, m_v = coefficient of volume decrease, k = permeability, and γ_w = unit weight of water.

Solving this equation will give the average degree of consolidation with time.

The classical theory described excess pore water pressure distribution as a function of

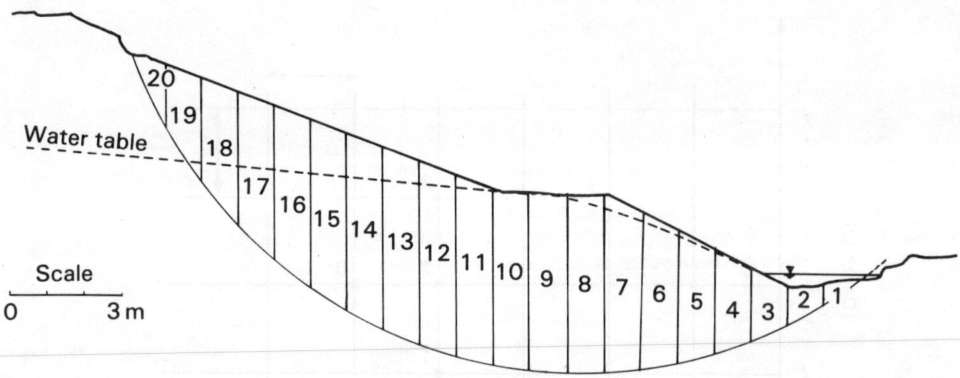

Slice No.	Width (m)	Height (m)	Angle (o)
1	1.0	0.4	−35
2	1.0	0.8	−27.5
3	1.0	1.5	−23
4	1.0	2.35	−18
5	1.0	3.1	−13.5
6	1.0	3.75	− 9.5
7	1.0	4.35	− 5.5
8	1.0	4.6	1.0
9	1.0	4.6	3.5
10	1.0	4.5	7.5
11	1.0	4.55	12.0
12	1.0	4.7	16.3
13	1.0	4.7	21
14	1.0	4.65	26
15	1.0	4.5	30.6
16	1.0	4.1	35.5
17	1.0	3.7	41.5
18	1.0	3.0	47
19	1.0	2.2	53.5
20	1.0	0.9	59.5

Fig. 11.9 Soil parameters: $\gamma = 20\ \text{kN/m}^3$, $r_u = 0.4$, $c' = ?$, $\phi' = ?$

depth and time. However, during consolidation as the excess pore water pressure dissipates, the soil undergoes strain, therefore, the consolidation equation can also be written in terms of strain, i.e.

$$C_v \frac{\partial^2 \epsilon}{\partial z^2} = \frac{\partial \epsilon}{\partial t} \tag{11.16}$$

where ϵ is strain in z direction.

The average degree of consolidation with time can be related to strain and the average strain in the clay layer will be equal to the degree of consolidation (U) multiplied by final strain (ϵ_f).

$$\epsilon_{av} = U\epsilon_f \tag{11.17}$$

The final strain is related to the total amount of consolidation settlement by Equation 11.18.

$$\epsilon_f = \Delta H/H \tag{11.18}$$

where ΔH = total consolidation settlement, and H = original thickness of the clay layer. Hence the average strain at time t after the application of a load on a layer of clay can be found from Equation 11.19.

$$\epsilon_{av} = \rho/H \tag{11.19}$$

where ρ = settlement at time t, and H = original thickness of clay layer.

Terzaghi's solution was analytical but the differential equation of consolidation can also be solved numerically by finite difference techniques. The numerical solution has the advantage in that it can conveniently and quickly accommodate stratified deposits in which C_v varies from stratum to stratum.

11.4.1 Numerical Solution

Let us consider the solution for the consolidation of a double-drained layer of clay. The solution must start with known initial values and changes from these values while always satisfying all the boundary conditions. The finite difference grid for the problem is shown in Fig. 11.10. The solution will

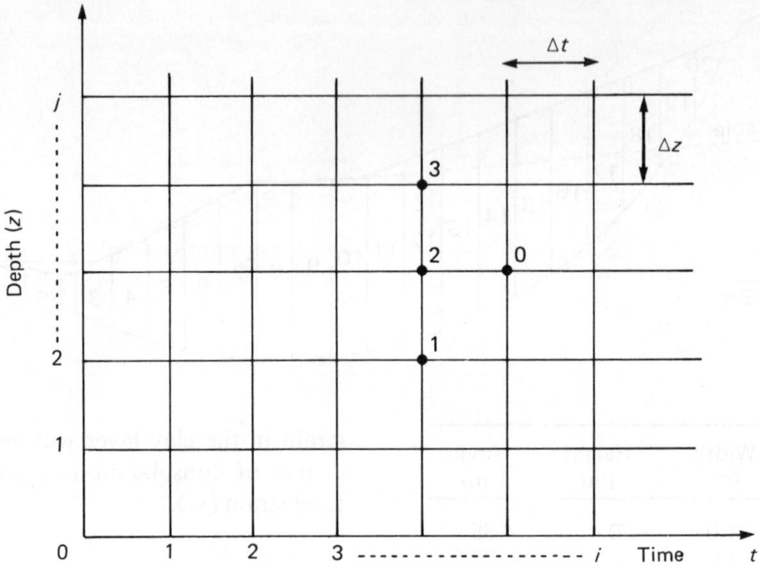

Fig. 11.10

advance from the z axis at time zero along the time axis.

With a grid size of Δt by Δz and by using the finite difference approximations for the consolidation equation, the strain at a particular point (i.e. point 0) can be obtained in terms of strain at the grid points 1, 2 and 3 of the previous time increment. It can, therefore, be seen that if the strains at a particular time increment are determined over a specific depth for the given boundary conditions then the strains for the next time increment can be solved in terms of those just obtained.

Let us assume points 0, 1, 2 and 3 have the following coordinates:

Point	Coordinates
0	i, j
1	$i-1, j-1$
2	$i-1, j$
3	$i-1, j+1$

The first derivative of strain (ϵ) with respect to time (t) will give us the forward difference approximation

$$\frac{\partial \epsilon}{\partial t_{i-1,j}} = \frac{\epsilon_{i,j} - \epsilon_{i-1,j}}{\Delta t} \tag{11.20}$$

The second derivative of strain (ϵ) with respect to depth z is used for the central difference approximation.

$$\frac{\partial^2 \epsilon}{\partial z^2_{i-1,j}} = \frac{\epsilon_{i-1,j+1} - 2\epsilon_{i-1,j} + \epsilon_{i-1,j-1}}{\Delta z^2} \tag{11.21}$$

The substitution of these equations in the differential consolidation equations yields the finite difference approximation

$$\frac{C_v}{\Delta z^2}(\epsilon_{i-1,j+1} - 2\epsilon_{i-1,j} + \epsilon_{i-1,j-1}) = \frac{\epsilon_{i,j} - \epsilon_{i-1,j}}{\Delta t} \tag{11.22}$$

Thus the strain at point 0 (i, j) is given by

$$\epsilon_{i,j} = \epsilon_{i-1,j} + \frac{C_v \Delta t}{\Delta z^2}(\epsilon_{i-1,j+1} - 2\epsilon_{i-1,j} + \epsilon_{i-1,j-1}) \tag{11.23}$$

In order to use this equation we must first define the problem and the boundary conditions for one-dimensional consolidation on a layer of clay with drainage at both boundaries (Fig. 11.11a).

The solution starts at t_0 (Fig. 11.11b) and the initial boundary conditions must be

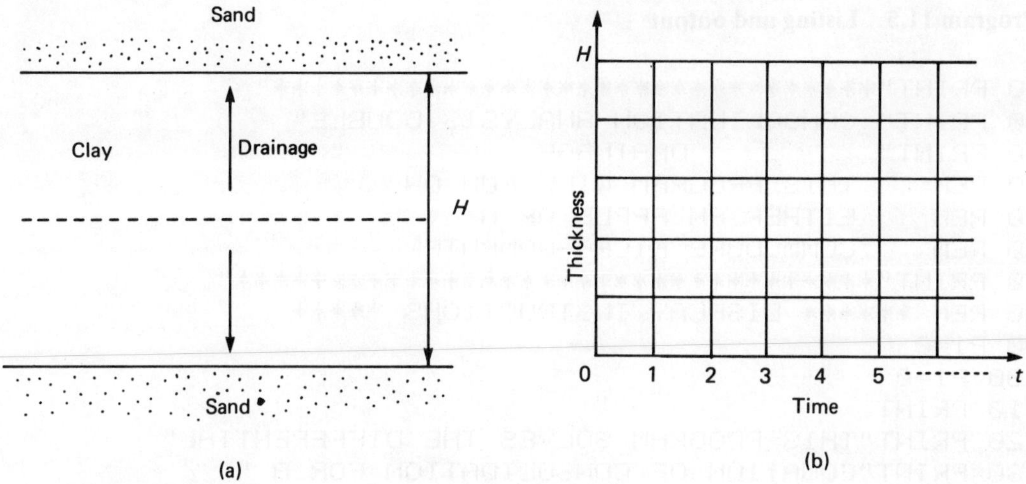

Fig. 11.11

known; also the boundary conditions along the boundaries $z = 0$ and $z = H$ with time must be defined. For double drainage these conditions are:

$$\epsilon = 0 \text{ at } t = 0$$

and at

$$z = 0 \text{ and } z = H, \text{ then } \epsilon = \epsilon_f$$

The strain at the top and bottom of the clay layer is final strain (i.e. 100% strain) due to the rapid drainage and compression which must occur at these boundaries. Thus the strain at every grid point can be determined and at any time the average strain as a percentage of total strain can be obtained by summation. The finite difference technique can cause truncation and round-off errors which can be minimized if

$$C_v \frac{\Delta t}{(\Delta z)^2} < 0.5$$

Program 11.5 calculates the average strain with time for a layer subjected to one-dimensional consolidation with double drainage. The user must supply the thickness of the clay layer, the coefficient of consolidation (in m²/day) and the time of interest for settlement. The depth increment Δz is also required but the user should ensure that the thickness of the clay layer divided by the depth increment gives an integer value. Settlement data are output as average strain, initially in 10 day intervals and then later in

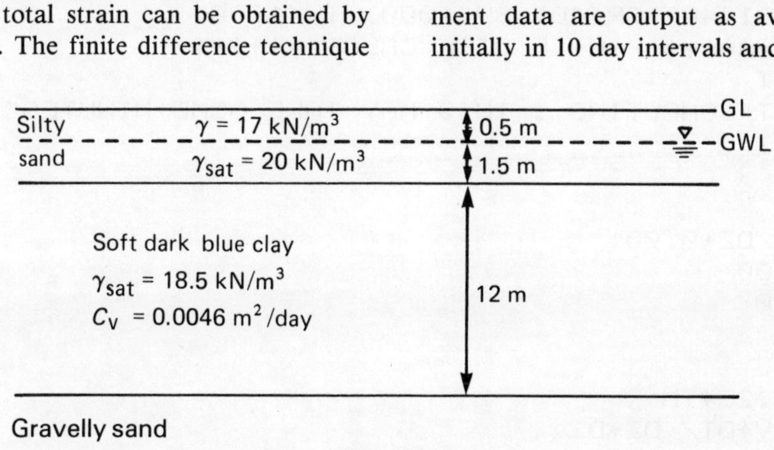

Fig. 11.12

Program 11.5 Listing and output

```
10 PRINT"************************************"
20 PRINT" CONSOLIDATION ANALYSIS DOUBLE"
30 PRINT"              DRAINAGE         "
40 REM     THIS PROGRAM WILL RUN ON
50 REM     EITHER AN APPLE OR A
60 REM     COMMODORE MICRO-COMPUTER
70 PRINT"**************************************"
80 REM ****** DISPLAY INSTRUCTIONS *****
90 P1=0
100 P1=0
110 PRINT
120 PRINT"THIS PROGRAM SOLVES THE DIFFERENTIAL"
130 PRINT"EQUATION OF CONSOLIDATION FOR A "
140 PRINT"LAYER OF CLAY WITH DOUBLE DRAINAGE."
150 PRINT"A DEPTH AND TIME INCREMENT ARE"
160 PRINT"SUPPLIED BY THE USER AND THE AVERAGE"
170 PRINT"STRAIN CALCULATED BY THE FINITE "
180 PRINT"DIFFERENCE METHOD"
190 PRINT
200 REM ****** INPUT DATA ***************
210 PRINT
220 PRINT"ENTER THE FOLLOWING DATA"
230 PRINT"THICKNESS OF CLAY LAYER(M)"
240 INPUT H
250 PRINT"DEPTH INCREMENT (M)      "
260 INPUT DZ
270 PRINT"COEFF. OF CONSOLIDATION(M↑2/DAY)"
280 INPUT CV
290 PRINT"TIME OF INTEREST (DAYS)"
300 INPUT TM
310 PRINT"TIME INCREMENT (DAYS)  "
320 INPUT DT
330 DIM S1(40),SR(40),SV(1000),TX(1000)
340 REM ****** FINITE DIFF/CALCS******
350 PRINT
360 PRINT"COMPUTING : THIS MAY TAKE SOME MINUTES"
370 PRINT
380 IT=0
390 KL=0
400 ND=H/DZ+0.001
410 T2=100
420 T1=10
430 T=0
440 TK=0
450 OH=0.25*TM
460 CT=CV*DT/(DZ*DZ)
470 SR(1)=100
```

```
480 S1(1)=100
490 SR(ND)=100
500 S1(ND)=100
510 FOR I=2 TO ND-1
520 S1(I)=0
530 NEXT I
540 T=T+DT
550 FOR I=2 TO ND-1
560 SR(I)=S1(I)+CT*(S1(I-1)-2*S1(I)+S1(I+1))
570 NEXT I
580 TK=TK+DT
590 IF T>OH THEN T1=T2
600 IF TK<T1 THEN GOTO 710
610 IT=IT+1
620 TX(IT)=T
630 REM **AVERAGE STRAIN***
640 SM=0
650 FOR J=1 TO ND
660 SM=SM+SR(J)
670 NEXT J
680 SV(IT)=SM/ND
690 TK=0
700 IF T>=TM THEN GOTO 760
710 FOR I=2 TO ND-1
720 S1(I)=SR(I)
730 NEXT I
740 T=T+DT
750 GOTO 550
760 PRINT
770 PRINT"=========================================="
780 PRINT"           CONSOLIDATION RESULTS"
790 PRINT"                DOUBLE DRAINAGE
800 PRINT"=========================================="
810 PRINT"  THICKNESS OF CLAY LAYER (M)";H
820 PRINT"  COEFF OF CONSOLIDATION(M↑2/DAY)";CV
830 PRINT"  DEPTH INCREMENT (M)";DZ
840 PRINT"  TIME PERIOD (DAYS) ";TM
850 PRINT"  TIME INCREMENT (DAYS)";DT
860 PRINT
870 PRINT" DAYS             AVERAGE STRAIN (%)"
880 PRINT
890 FOR I=1 TO IT
900 PRINT TX(I);TAB(20);SV(I)
910 NEXT I
920 PRINT
930 END
READY.
```

```
************************************
 CONSOLIDATION ANALYSIS DOUBLE
          DRAINAGE
**************************************

THIS PROGRAM SOLVES THE DIFFERENTIAL
EQUATION OF CONSOLIDATION FOR A
LAYER OF CLAY WITH DOUBLE DRAINAGE.
A DEPTH AND TIME INCREMENT ARE
SUPPLIED BY THE USER AND THE AVERAGE
STRAIN CALCULATED BY THE FINITE
DIFFERENCE METHOD


ENTER THE FOLLOWING DATA
THICKNESS OF CLAY LAYER(M)
 12
DEPTH INCREMENT (M)
 1
COEFF. OF CONSOLIDATION(M↑2/DAY)
 4.63E-03
TIME OF INTEREST (DAYS)
 1860
TIME INCREMENT (DAYS)
 1

COMPUTING : THIS MAY TAKE SOME MINUTES


========================================
        CONSOLIDATION RESULTS
            DOUBLE DRAINAGE
========================================
  THICKNESS OF CLAY LAYER (M) 12
  COEFF OF CONSOLIDATION(M↑2/DAY) 4.63E-03
  DEPTH INCREMENT (M) 1
  TIME PERIOD (DAYS) 1860
  TIME INCREMENT (DAYS) 1

 DAYS              AVERAGE STRAIN (%)

 10                        17.421193
 20                        18.144198
 30                        18.837129
 40                        19.5025179
 50                        20.1426288
 60                        20.75949
 70                        21.3549213
 80                        21.9305587
```

90	22.4878754
100	23.0282003
110	23.5527343
120	24.0625647
130	24.5586773
140	25.0419681
150	25.5132523
160	25.9732733
170	26.4227099
180	26.8621835
190	27.2922633
200	27.7134721
210	28.1262906
220	28.5311616
230	28.9284939
240	29.3186651
250	29.7020247
260	30.078897
270	30.4495829
280	30.8143624
290	31.1734961
300	31.5272274
310	31.8757832
320	32.219376
330	32.5582048
340	32.8924563
350	33.2223055
360	33.5479172
370	33.8694466
380	34.1870396
390	34.5008342
400	34.8109606
410	35.1175419
420	35.4206947
430	35.7205294
440	36.0171506
450	36.3106579
460	36.6011456
560	39.357767
660	41.8893935
760	44.2424922
860	46.4489005
960	48.5313047
1060	50.5063564
1160	52.3865779
1260	54.1815911
1360	55.8989433
1460	57.5446774
1560	59.1237337

```
1660                                    60.6402356
1760                                    62.0976962
1860                                    63.4991672


READY.
```

100 day intervals.

The program can be tested using the problem data given in Fig. 11.12 and the results compared with those of the trial run in Program 11.5.

Exercises

11.1 Modify the liquid limit program (11.1) to incorporate graphical output to either VDU or line printer.

11.2 Rewrite the appropriate part of Program 11.4 (slope stability) to allow repeat runs of the program for simple back-analysis. This can be achieved by incorporating a routine at the end of the program allowing c' and ϕ' values to be changed before returning control to line 540 for the calculation of a factor of safety.

11.3 Determine the values of c' and ϕ' for the example in Fig. 11.9 at failure using the modified program from Question 11.2.

11.4 If the pore water pressure ratio of the slope in Fig. 11.8 increases to 0.5 obtain the new factor of safety of the slope.

11.5 Modify Program 11.5 (consolidation analysis) to request the total expected settlement of the clay layer due to a surface load and output the amount of settlement with time.

11.6 If the total settlement of the clay layer in Fig. 11.12 is 189 mm due to a surface load plot the time/settlement curve over a five year period. Use the modified program from Question 11.5.

Further Reading

Bishop, A. W. 'The use of the slip circle in the stability analysis of earth slopes.' Geotechnique, **5**, pp. 1–17, 1955.

British Standard 1377. '1975 Methods of test for soil for civil engineering purposes.' British Standards Institution, London.

CP 110. '1972 Code of Practice for the structural use of concrete.' British Standards Institution, London.

Franklin, J. A. 'Triaxial strength of rock materials.' *Rock Mechanics*, **3**, pp. 86–98, 1971.

Hubert, M. K. and Rubey, W. W. 'Role of fluid pressure in mechanics of overthrust faulting.' *BGSA*, **70**, p. 115, 1959.

Lisle, R. J. and Strom, C. S. 'Least-squares fitting of the linear Mohr envelope. *Q. J. Eng. Geol. London*, **15**, pp. 55–56, 1982.

12

Interfacing

12.1 Introduction

The sharply rising cost of manpower and the declining cost of electronic components has resulted in automation in many labour intensive industries. This trend towards automation is very evident in engineering laboratories and with the help of microcomputers data collection and processing can be improved. For certain tasks, computers can be programmed to start a test, collect data, stop the test and analyse and display the results. By means of feedback loops they can assess how the test is progressing and vary the test procedure accordingly. Long-term testing is no problem and in some instances one microcomputer can monitor several pieces of equipment carrying out different tests. The level of automation will depend not only on the type of microcomputer but also, and more importantly, on the software used to control the testing procedures. When a microcomputer is used to monitor and control tests it must be capable of transferring and receiving data from test equipment and measuring devices (transducers and sensors).

12.2 Data Transfer and Communication

Connection to printers, disk units and test equipment to achieve communication is known as 'interfacing'. By definition an interface is a common boundary between two regions, thus, when a computer is connected to a printer there must be a common method between the two pieces of equipment enabling data to be transferred from one to the other.

But how does a computer transmit information? Chapter 3 briefly explained how information was stored in bytes which consisted of a number of bits and it is in this form that data are transferred. A byte will represent a character which will have a specific meaning to the piece of equipment receiving it. Thus bytes must be transmitted along the wires which connect the microcomputer to the receiving equipment (receiver). The wires along which data are passed constitute a *bus* and, just as road traffic must obey the Highway Code for safe and steady movement, so the movement along *buses* must follow rules. The rules for data transference are referred to as *protocol*.

The speed at which information is transmitted is assessed as the number of bits per second and is called the *baud rate*. Bits per second are used to measure the rate as some machines use 16 bit words and larger. When 8 bits are used a rough estimate of the number of characters that can be transmitted or received can be obtained by dividing the baud rate by ten.

The transference of information can be accomplished using either a *serial* bus or a *parallel* bus.

12.3 Serial Bus

This is the simplest form and it is referred to as a *data link* rather than a bus. One line or wire is used and a byte is transmitted one bit at a time. They are subdivided into 'asynchronous' or 'synchronous' data links.

Asynchronous serial data links operate on groups of bits representing characters coded in the ASCII code. Because the bytes are transferred in groups of bits the receiver must have a method of distinguishing between one group and another and between individual bits. This can only be done by separating information by time. When a line is inactive it will be in what is known as a 'logic state zero' and, therefore, can represent a bit 0. To indicate a bit 1 the level has to be changed to represent a 'logic state one'. Some interfaces use a particular amount of current as the zero state, others use the more common method of a reference voltage as the zero state. When voltage is used the level one state is simply represented by an increase in the voltage level (Fig. 12.1).

When a microcomputer is about to send a character (a byte) it must first issue a control telling the receiver that information is coming. It does this by raising the level from zero to one. The receiver as soon as it detects this, starts its own internal clock and waits for T seconds after which if the level is still state one it samples the data (logic states) on the line at a specific time interval T_1 (where $T_1 \neq T$), until it receives the last bit in the group. It then switches off the clock and waits for the next signal to say that another group of bits is on the line. If, after T seconds, the logic state had returned to zero the receiver would have done nothing. It is important that the micro-

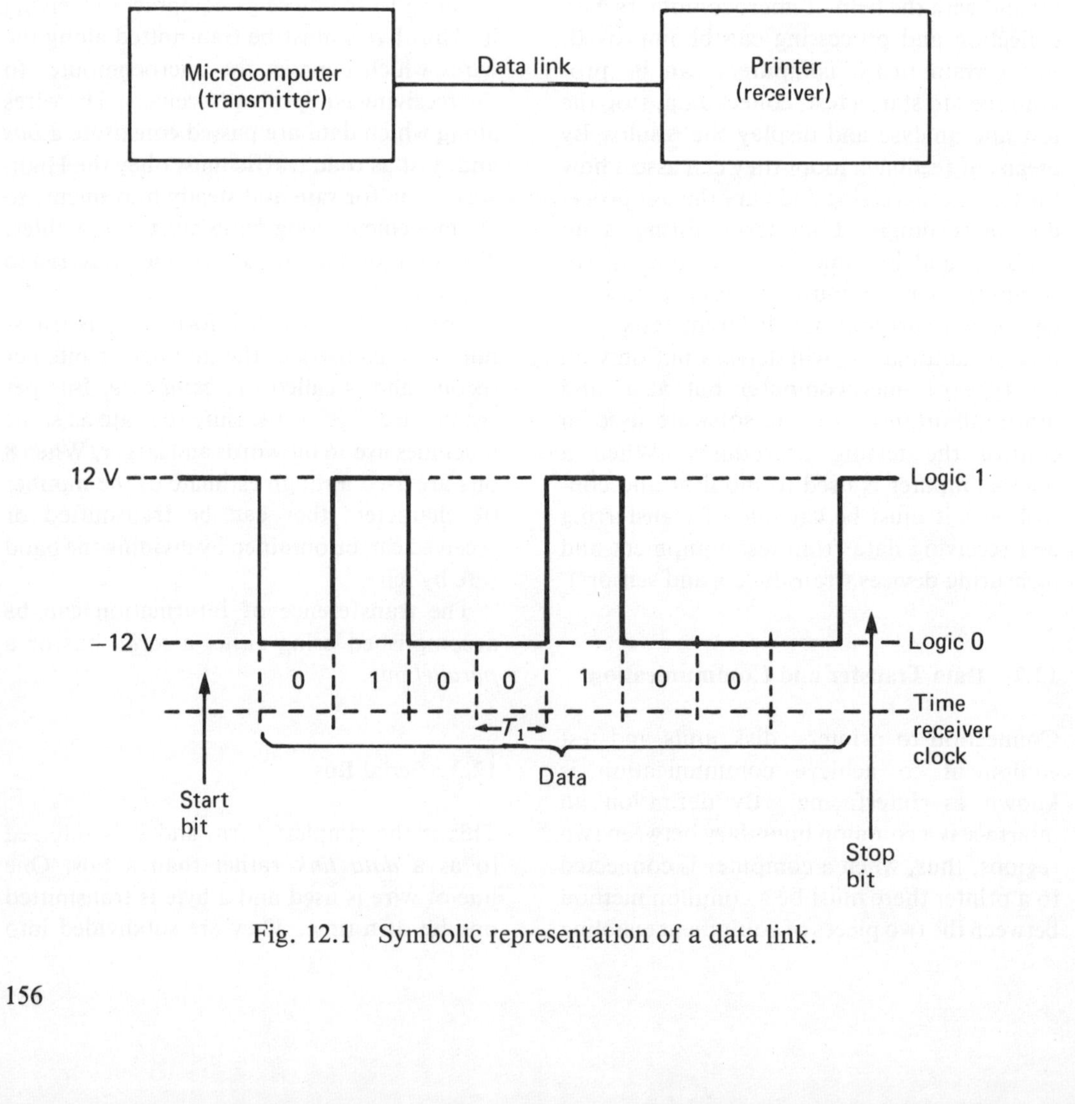

Fig. 12.1 Symbolic representation of a data link.

computer (transmitter) puts the bits on the data line at the same rate at which the receiver can sample as the clock in the receiver is *not* synchronized with the transmitter, thus they must transmit and receive at the *same* baud rate.

Synchronous data links have the clocks of the transmitter and receiver synchronized hence it is unnecessary to group the bits together into bytes. The receiver will receive a continuous string of bits from which it can automatically reconstruct the characters.

Because only a single line is used this method is cheaper than a parallel bus and is used when economy is more important than speed. The RS 232 and the 20 mA loop are the common data links.

12.4 Parallel Bus

Parallel buses move a number of bits simultaneously along parallel wires using one wire per bit and can be classed as either *open-loop* or *closed-loop* systems (Fig. 12.2).

In the open-loop system, the transmitter sends the data and, after a suitable time delay, automatically assumes that the data have been received. It is then free to send along another byte. The closed-loop system necessitates *two-way* communication between the transmitter and receiver. Any data sent must be acknowledged by the receiver before any more are transmitted. The closed system is used to its best advantage with 'intelligent' peripherals. The number of wires required are one for each bit plus some control lines (Fig. 12.2). For example, the receiving device must know when data are to be transmitted; this signal is sent along a control line. The receiver acknowledges receipt via a different control line. This interaction between transmitter and receiver is called 'handshaking' and can use either two or three control lines.

The IEEE-488 General Purpose Interface Bus (GPIB) is one of the three line types and is widely used as a standard with intelligent test equipment. The IEEE-488 was originally proposed by Hewlett Packard and was later adopted by the Institution of Electrical and Electronic Engineers. This interface allows up to 15 instruments to be connected on the same cable and is the main interface on Commodore and Hewlett Packard computers. The protocol required with this interface is relatively easy to follow but is beyond the scope of this text. The use of the Commodore microcomputer with devices which use the IEEE-488 simplifies the programming as some of the protocol is taken care of when addressing the devices using the OPEN and PRINT statements. The reason for this is that Commodore microcomputers use the IEEE-488 to communicate with printers and disk units and, therefore, protocol must be automatically taken care of for simplicity of use.

12.5 Measuring Devices

Testing and monitoring equipment obtain

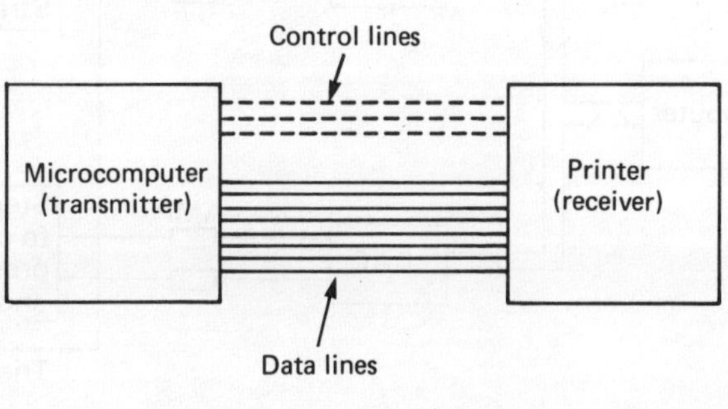

Fig. 12.2 Parallel bus.

data not in bits but in some other form, frequently, voltage. This must be converted to a series of bits prior to its being sent to a microcomputer. In civil engineering the most common measurements are associated with stress and strain. These data can be measured via dial gauges which monitor displacements (length), which can be converted to stress or strain by factors. For example, the shortening diameter of a proving ring can be converted to a load using the proving ring factor and, in the case of a triaxial soil test, it can be converted to stress if the cross-sectional area of the soil sample is known.

Displacement can be measured electronically by using LVDT's ('Linearly Variable Differential Transformers'). Transducers can be of a variety of types and design, but basically they convert a physical variable to an electrical signal (usually voltage). The foil strain gauges work in a similar way. The voltage (or signal) obtained from the transducer normally has to be amplified before it can be used by a display system.

A simple display system would be the conversion of the signal directly to a digital display as found on digital voltmeters. A computer cannot understand direct values of voltage and so another unit is required to convert the transducer's signal into a series of 0's and 1's (bits), in other words the digital equivalent of the actual voltage measured. The unit that converts the voltage is known as an 'Analogue to Digital Convertor' (ADC) and these are normally constructed with one of the communications interfaces (Sections 12.3 and 12.4) built in.

A computer can also transmit data and, therefore, it can control test equipment. However, test equipment needs analogue signals and the digital control data supplied by a computer must be converted by a 'Digital to Analogue Convertor' (DAC) before it reaches the test equipment. Such an analogue signal could be used to control motors, in particular the stepper motor (Fig. 12.3).

An example of an intelligent pressure and volume measuring device used in soil mechanics is that manufactured by GDS Triaxial Systems Ltd. The device is a pressure controller for a triaxial cell which can be used to monitor volume changes. It has its own microprocessor which can work independently of a microcomputer, but by means of its IEEE-488 interface it can also be operated under program control allowing sophisticated stress path triaxial testing to be accomplished with ease.

12.6 Software

The advantage of using a microcomputer with monitoring equipment in a laboratory is

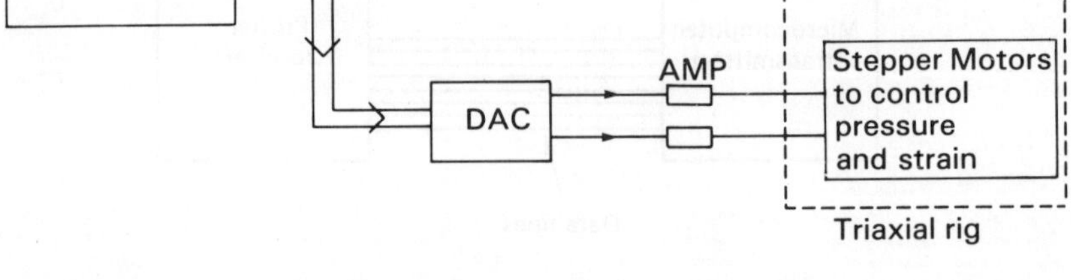

Fig. 12.3

its versatility, since it can be programmed to monitor a variety of tests and analyse the results. The main problem, as far as this text is concerned, is that software will be dependent not only on the computer but also on the communications interface and the monitoring device.

Monitoring tests can, however, be controlled from a BASIC program but in certain applications the speed of sampling may not be fast enough which will require part of the program to be written in machine code.

There is one feature on many microcomputers which facilitates sampling and control and that is the computer's internal clock. This facility varies from computer to computer and only the Commodore version will be described here.

The Commodore 4000 and 8000 series computer internal clock can be accessed by the reserved variables *TI$* and *TI*. The numeric variable *TI* is essentially a counter which is increased by 1 every 1/60th of a second. Commodore call this unit a 'jiffy', thus 1 second consists of 60 jiffies.

Every time a call is made to access *TI$*, the numerical variable *TI* is converted to hours, minutes and seconds, the result being stored in the variable *TI$* in the following format

TI$ = "HHMMSS"

where HH is the hours; MM the minutes and SS seconds.

When a Commodore computer is switched on the clock starts counting immediately from zero and therefore when *TI$* is printed out it will give elapsed time from switch on. After 24 hours it automatically resets itself when *TI* reaches 5 183 999 jiffies. It is possible to reset the time variables at any time to show real time, for example

TI$ = "104010"

will reset the time to 10–40 a.m. and 10 seconds. When *TI$* is reset the *TI* variable is also adjusted to the correct number of jiffies. Both of these variables can be used for timing tests and monitoring test equipment at specific times.

Program 12.1 demonstrates the use of the clock in a Commodore 4000 series computer to monitor one load cycle of an oedometer test. The program was written for an ADC with a IEEE-488 interface. The sampling routine using this device would be placed at lines 246Ø to 256Ø, however, this has been left blank apart from instructions for a user to put in his own routine to match his own devices.

The program uses a constant VDU display showing the elapsed time from the start of the test in digital clock format (Fig. 12.4). The clock is updated every second and compression of the oedometer sample is displayed on a line printer. The program is designed around a main sampling routine with subroutines being called when required. The program is structured as follows:

Lines 19Ø–23Ø The variables *A$* and *C$* are dimensioned and initialized. These variables are used in the main sampling routine.

Lines 24Ø–26Ø The input routines are requested; the first subroutine 164Ø–17ØØ reads the time values into *AL$*. Each element of *AL$* represents a time when sampling the oedometer must take place according to the BS 1377 Soil Testing Standard. Subroutine 177Ø–185Ø then requests specific data about the sample being tested. The transducer coefficient is a factor used to convert the transducer readings to millimetres.

Lines 27Ø–3ØØ These lines call the routines that select and open the communica-

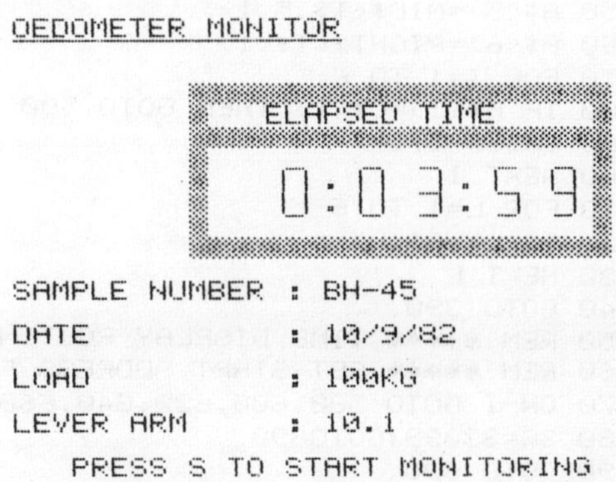

Fig. 12.4 Program 12.1 screen display.

Program 12.1 Listing

```
10 REM ********************************
20 REM    SINGLE OEDOMETER MONITOR
30 REM    COMMODORE VERSION BASIC 2 & 4
40 REM   CURSOR CONTROL CHARACTERS ARE
50 REM    SHOWN  AS FOLLOWS :
60 REM       [CLR] : CLEAR SCREEN
70 REM       [HOM] : HOME CURSOR
80 REM       [CD]  : CURSOR DOWN
90 REM       [CU]  : CURSOR UP
100 REM      [CR]  : CURSOR RIGHT
110 REM      [CL]  : CURSOR LEFT
120 REM      [RVO] : REVERSE VIDEO ON
130 REM      [RVF] : REVERSE VIDEO OFF
140 REM  TO INDICATE MORE THAN ONE OF
150 REM  THE ABOVE A NUMBER IS PLACED
160 REM  WITHIN THE BRACKETS.
170 REM      [6CD]
180 REM ********************************
190 PRINT"[CLR]"
200 DIM A$(6),C$(6)
210 FOR I=1 TO 6
220 C$="0"
230 NEXT I
240 REM ***** INPUT ROUTINE ************
250 GOSUB 1640
260 GOSUB 1770
270 REM  **** OUTPUT INITIAL DATA *****
280 GOSUB 1860
290 REM **** SET UP SCREEN DISPLAY *****
300 GOSUB 2130
310 TI$="000000"
320 GOSUB 2460
330 D1=D2
340 SR=D2
350 GETA$
360 IF A$<>"S" THEN 350
370 TI$="000000"
380 REM *** SAMPLE TIME MAIN ROUTINE ***
390 IF TI$=>AL$(AL) THEN GOSUB 2460:GOSUB 2570
400 T$=TI$
410 A$(1)=LEFT$(T$,1)
420 A$(2)=MID$(T$,2,1)
430 A$(3)=MID$(T$,3,1)
440 A$(4)=MID$(T$,4,1)
450 A$(5)=MID$(T$,5,1)
460 A$(6)=RIGHT$(T$,1)
470 FOR I=1 TO 6
480 IF A$(I)=C$(I) THEN GOTO 500
490 GOSUB 570
500 NEXT I
510 FOR L=1 TO 6
520 C$(L)=A$(L)
530 NEXT L
540 GOTO 390
550 REM ***** TIME DISPLAY ROUTINE ****
560 REM ***** SET START ADDRESS FOR DIGIT *****
570 ON I GOTO 580,600,620,640,660,680
580 SA=33099:GOTO690
590 GOTO 690
600 SA=33102:GOTO690
610 GOTO 690
```

```
620 SA=33106:GOTO690
630 GOTO 690
640 SA=33109
650 GOTO 690
660 SA=33113:GOTO690
670 GOTO 690
680 SA=33116
690 NI=VAL(A$(I))
700 IF NI=0 THEN IFI=1 THEN GOSUB 1570
710 IF NI=0 THEN IFI=1 THEN RETURN
720 IF NI=0 THEN GOSUB 1480
730 IF NI=0 THEN RETURN
740 REM ***** DIGIT DISPLAY SUBROUTINES *****
750 ON NI GOSUB 780,830,910,990,1060,1140,1230,1290,1390,1480
760 RETURN
770 REM ***** ONE ********************
780 GOSUB 1570
790 POKE SA+41,101
800 POKE SA+81,101
810 RETURN
820 REM ***** TWO    ******************
830 GOSUB 1570
840 POKE SA,100
850 POKE SA+41,101
860 POKE SA+80,99
870 POKE SA+120,99
880 POKE SA+79,103
890 RETURN
900 REM ***** THREE ******************
910 GOSUB 1570
920 POKE SA,100
930 POKE SA+41,101
940 POKE SA+80,99
950 POKE SA+81,101
960 POKE SA+120,99
970 RETURN
980 REM ***** FOUR   ******************
990 GOSUB 1570
1000 POKE SA+39,103
1010 POKE SA+41,101
1020 POKE SA+81,101
1030 POKE SA+80,99
1040 RETURN
1050 REM ***** FIVE   ******************
1060 GOSUB 1570
1070 POKE SA,100
1080 POKE SA+39,103
1090 POKE SA+80,99
1100 POKE SA+81,101
1110 POKE SA+120,99
1120 RETURN
1130 REM ***** SIX    ******************
1140 GOSUB 1570
1150 POKE SA,100
1160 POKE SA+39,103
1170 POKE SA+79,103
1180 POKE SA+81,101
1190 POKE SA+120,99
1200 POKE SA+80,99
1210 RETURN
```

```
1220 REM ***** SEVEN ******************
1230 GOSUB 1570
1240 POKE SA,100
1250 POKE SA+41,101
1260 POKE SA+81,101
1270 RETURN
1280 REM ***** EIGHT ******************
1290 GOSUB 1570
1300 POKE SA,100
1310 POKE SA+39,103
1320 POKE SA+41,101
1330 POKE SA+80,99
1340 POKE SA+81,101
1350 POKE SA+120,99
1360 POKE SA+79,103
1370 RETURN
1380 REM ***** NINE  ******************
1390 GOSUB 1570
1400 POKE SA,100
1410 POKE SA+39,103
1420 POKE SA+41,101
1430 POKE SA+81,101
1440 POKE SA+120,99
1450 POKE SA+80,99
1460 RETURN
1470 REM ***** ZERO  ******************
1480 GOSUB 1570
1490 POKE SA,100
1500 POKE SA+39,103
1510 POKE SA+41,101
1520 POKE SA+79,103
1530 POKE SA+81,101
1540 POKE SA+120,99
1550 RETURN
1560 REM ***** BLANK ******************
1570 FOR J=SA TO SA+120 STEP 40
1580 POKE J,CS
1590 POKE J+1,CS
1600 POKE J-1,CS
1610 NEXT J
1620 RETURN
1630 REM ***** END OF PROGRAM **********
1640 DIM AL$(49)
1650 FOR I=1 TO 49
1660 READ AL$(I)
1670 NEXT I
1680 AL=1
1690 AC=40
1700 RETURN
1710 DATA 000000,000005,000010,000015,000020
1720 DATA000030,000045,000060,000215,000400,000615,000900
1730 DATA 001215,001600,002015,002500,003600,004900,010400,012100,014000,02010
1740 DATA022400,024900,031600,034500,041600,044900,052400,060100,064000
1750 DATA072100,080400,084900,093600,102500,111600,120900,130400,140100,150000
1760 DATA160100,170400,180900,191600,202500,213600,224900,235958
1770 REM ****  INPUT ROUTINE *****
1780 PRINT "[CLR][2CD]ENTER THE FOLLOWING DATA"
1790 INPUT "[CSD]DATE          :";D$
1800 INPUT "[CSD]SAMPLE NUMBER :";SN$
1810 INPUT "[CSD]OPERATOR      :";O$
1820 INPUT "[CSD]TRANSDUCER COEFF (MM/UNIT)";TC
```

```
1830 INPUT "[CSD]LOAD            :";L$
1840 INPUT "[CSD]LEVER ARM RATIO :";AR$
1850 RETURN
1860 REM ***** PRINTER TITLE ROUTINE ****
1870 OPEN 1,4
1880 OPEN 2,4,2
1890 OPEN 3,4,1
1900 CR$=CHR$(29)
1910 PRINT#2,"AAAAAAAAAA    AAAAAAAAAA    AAAAAAAA  AAAAAAAA  AAAAAAAA"
1920 FOR I=1 TO 78
1930 PRINT#1,"-";
1940 NEXT I
1950 PRINT#1,CHR$(1),CHR$(1);"OEDOMETER TEST"
1960 FOR I=1 TO 79
1970 PRINT#1,"=";
1980 NEXT I
1990 PRINT#1
2000 PRINT#1," DATE     .......................";D$
2010 PRINT#1
2020 PRINT#1," SAMPLE NUMBER      ............";SN$
2030 PRINT#1
2040 PRINT#1," OPERATOR       ................";O$
2050 PRINT#1
2060 PRINT#1," TRANSDUCER COEFF (MM/UNIT).";TC
2070 PRINT#1
2080 PRINT#1," LEVER ARM RATIO ............";AR$
2090 PRINT#1
2100 PRINT#1,"  TIME          RESULT      DIFF        CUM        CUM COMP(MM)
2110 PRINT#1,"HHMMSS           "
2120 RETURN
2130 REM ****** DRAW CLOCK OUTLINE ******
2140 CS=96
2150 CH=230
2160 PRINT "[CLR]"
2170 SA=33057
2180 SE=33078
2190 FOR I=32977 TO 32998
2200 POKE I,CH
2210 POKE I+40,CH
2220 NEXTI
2230 PRINT"[HOM][6CD][10CR][RVO]   ELAPSED TIME    "
2240 PRINT"[HOM]"
2250 FOR I=SA TO SE
2260 POKE I,CH
2270 POKE I+200,CH
2280 NEXT I
2290 FOR I=SA+40 TO SA+160 STEP 40
2300 POKE I,CH
2310 POKE I+21,CH
2320 NEXT I
2330 POKE 33144,46
2340 POKE 33151,46
2350 POKE 33184,46
2360 POKE 33191,46
2370 PRINT"[HOM][2CD]OEDOMETER MONITOR"
2380 PRINT"_________________"
2390 PRINT"[10CD]SAMPLE NUMBER : ";SN$
2400 PRINT"[CD]DATE          : ";D$
2410 PRINT"[CD]LOAD          : ";L$
2420 PRINT"[CD]LEVER ARM     : ";AR$
2430 PRINT"[CD][RVO]   PRESS S TO START MONITORING   "
```

```
2440 PRINT"[HOM]"
2450 RETURN
2460 REM **** SAMPLE ROUTINE **********
2470 REM  THE USER SHOULD INSERT IN THE
2480 REM  FOLLOWING LINES THE ROUTINE
2490 REM  WHICH SAMPLES DATA VIA HIS
2500 REM  INTERFACE UNIT
2510 REM  THE DATA SHOULD BE CONVERTED
2520 REM  TO A DISPLACEMENT VALUE USING
2530 REM  THE TRANSDUCER COEFFICIENT AND
2540 REM  THE VARIABLE D2 SET TO THIS
2550 REM  VALUE.
2560 REM ******************************   2516RETURN
2570 REM ***** PRINTER OUTPUT ROUTINE ***
2580 RD=D2-D1
2590 CD=D2-SR
2600 CP=CD*TC
2610 RD$=STR$(RD)
2620 CD$=STR$(CD)
2630 CP$=STR$(CP)
2640 PRINT#3,AL$(AL),CR$,R$,CR$,RD$,CR$,CD$,CR$,CP$
2650 AL=AL+1
2660 IF AL<49 THEN RETURN
2670 D1=D2
2680 SR=D2
2690 RETURN
2700 CLOSE1:CLOSE2:CLOSE3
READY.
```

tion channel to the printer and then prints out the headings for the data to be collected. Subroutine 213Ø to 245Ø sets up the screen display prior to the start of the test, informing the user that when everything is ready the monitoring can be started by pressing the 'S' key. This must be done at the same time as the load increment is applied to the sample. If a sound box is available the routine could be rewritten to signal the user by sound when to place the load on the sample.

Lines 39Ø–54Ø Prior to pressing the "S" key the computer will have taken the first set of readings when the elapsed time is zero. When the "S" key is pressed it enters into the lines 39Ø–54Ø and it will then loop round these lines continuously only jumping out temporarily when the clock display has to be changed or when data are to be obtained from the oedometer. The decision to sample is taken at line 39Ø where the sampling times in *AL$* are checked against the elapsed time. Only one element is checked and this depends on the value of *AL*. The variable *AL* starts at a value of 1 and is only increased by one after sampling, in this way the time at which sampling takes place changes according to the values read into the array *AL$*.

When a sample has to be taken, control jumps to routine 246Ø and then prints out the result by jumping to subroutine 257Ø.

Lines 4ØØ–54Ø These lines check to see if any part of the time display HH, MM or SS has changed since it previously checked. If any has changed lines 55Ø to 163Ø are the routines which change the numbers on the screen clock.

The sample program briefly described only

monitors one oedometer and uses the variable *TI$*. A more efficient program can be achieved by using the numeric variable *TI*, omitting the digital time display and thus allowing the program to run faster and monitor eight or more oedometers simultaneously.

The crucial importance of the software in the successful recording and processing of laboratory data should be evident from this illustration.

Further Reading

Fisher, E. and Jensen, C. W. *Pet and the IEEE 488 Bus (GPIB)*. McGraw-Hill, 1981.

Leventhal, L. A. *Introduction to Microprocessors*. Prentice/Hall International Editions, 1978.

Appendix 1

Computer Vocabulary

Accumulator A memory location used by the Central Processing Unit to manipulate data.

Add The addition of two numbers using the accumulator.

Address A number that identifies a memory location in which a binary number can be stored.

Algol A high level programming language.

AND A logical operation.

Architecture The arrangement of the hardware and internal software that make up a microcomputer.

ASCII A set of characters based on an 8 bit pattern, American Standard Code for Information Interchange.

Assembler A computer program that translates symbolic machine code to binary machine code.

Assembly Language A symbolic code used to represent the binary machine code.

BASIC A high level programming language.

Baud Rate The rate at which data can be transmitted in serial connection between computer and peripheral equipment. Measured in bits per second.

BCD Binary Coded Decimal. A code used with microprocessors and calculators in which each digit of a decimal number is represented by 4 bits, e.g. decimal 12 in BCD is

```
   1   :  2
       :
0001   :  0010
       :
```

Binary Digit Only two digits are possible, i.e. 0 or 1.

Bit A single binary digit.

Breakpoint The position in a program at which execution is stopped.

Buffer A series of memory locations used as temporary stores during data handling. Buffers are normally associated with input/output operations from peripherals.

Bug A mistake in a program, also used to refer to a hardware fault.

Bus A series of wires that carry parallel binary data.

Byte A number of bits, commonly 8 bits.

Central Processor Unit Central Processor Unit (CPU) which executes the instructions of a program and controls all operations.

Character Set A set of characters which includes alphabetic letters, numbers, etc.

Clock A regular time signal that controls the operations of the CPU.

Cobol A high level language that was written for business usage.

Code Symbols in the form of a language that is understandable to the computer system.

Compiler A program which translates instructions written in a high level language to machine code understandable by the Central Processor Unit.

CPU See Central Processor Unit.

CRT Cathode Ray Tube.

Data Store Any medium for storing data.

Debug To remove errors from a program.

Dedicated Refers to a system or a device used exclusively for one application.

Disk Operating System (DOS) An operating system that controls the transference of data and programs to and from disk units.

Diskette See Floppy Disk.

EPROM Erasable Programmable Read Only Memory. Erasure is carried out using ultraviolet light.

Execute Execution of program or instructions.

Execution Time Time taken for program or an instruction to be completed.

Firmware Programs permanently stored in read-only memories for system use.

Floppy Disk A circular flexible magnetic surface used for data storage. Also referred to as a Diskette.

Flow Chart A graphical representation of the procedure followed in a computer program.

FORTRAN A high level programming language devised for scientific problem solving.

Hardware Equipment that constitutes a computer system.

High Level Language A programming language such as COBOL, FORTRAN and BASIC which is easy to use and understand but requires a compiler or interpreter to convert it to machine code.

Input/Output Ports Electrical terminals used by the computer to communicate with the outside world.

Instructions A group of bits that defines a single instruction to the computer.

Interface A device that enables a computer to communicate with other devices.

Interpreter A program that fetches and executes statements written in a high level language. It does not produce an object program as a compiler does.

Language A series of statements used to communicate data and instructions to the computer.

Library A collection of standard programs and routines accessible to a computer user.

Machine Code A programming language that a computer can directly understand.

Memory The part of the computer that stores the data and program instructions. Each memory location has a unique address and can hold one computer word, in the case of the majority of microcomputers one byte.

Mnemonic An easily remembered substitute for a word or phrase.

Page A subdivision of the memory section.

Peripheral A piece of hardware that is connected to a computer.

Ports An addressable input/output memory area.

Program A sequence of properly ordered instructions that perform a specific task.

PROM Programmable Read Only Memory.

RAM Random Access Memory, the memory locations can be read from or written to by a program.

ROM Read Only Memory, memory locations that can be read but *not* written to. Data are fixed in these locations during manufacture.

RS-232 A standard interface for the transmission of serial digital data.

Software See 'program'.

Syntax The grammar of a programming language.

User Person who operates a computer.

VDU Visual Display Unit. Data are displayed as characters on a TV-type screen.

Volatile Memory This is RAM that loses the stored information when power is removed.

Word The basic grouping of bits that a computer can operate on in a single operation. Word lengths in microcomputers are either one or two bytes long.

Appendix 2

Summary of BASIC

Variables

Type	Name
Real	*AC*
Integer	*AC%*
String	*AC$*

Where *A* is a letter, *C* is a letter or digit.

Array Variables

Type	Name
Real	*AC* (*X*, *Y*, *Z*)
Integer	*AC%* (*X*, *Y*, *Z*)
String	*AC$* (*X*, *Y*, *Z*)

Where *X*, *Y*, *Z* are the subscripts defining an element of the array.

Algebraic Operators

- − Subtraction
- \+ Addition
- / Division
- * Multiplication
- ↑ Exponentiation
- = Assign

Relational Operators

=	Equal
< >	Not equal
<	Less than
>	Greater than
< =	Less than or equal
> =	Greater than or equal

System and Editing Commands

RUN	Execute the current program from the lowest line number.
NEW	Clears current program and data out of memory.
LOAD	Loads program from backing store.
SAVE	Saves current program to backing store.
STOP	Halts execution from a program and indicates in which line.
END	Terminates execution of a program (with no message).
CONT	Continues program execution that has been stopped by STOP or END.
LIST	Lists an entire program. Use of parameters enables parts of programs.
CLR	Reset all variables to zero.
PEEK (*X*)	Examines memory location *X*.
POKE *X*, *Y*	Stores the value *Y* in memory location *X*.

Input/Output Commands and Functions

INPUT	Requests data during program execution.
READ	Assigns data read from a DATA statement to variable(s).
DATA	Establishes a list of data that can be used by READ statements.
RESTORE	Resets the pointer to the begin-

	ning of the DATA list to enable the data to be read again.
PRINT	Outputs information to main output device (i.e. screen or printer).
GET	Accepts a single character or number from the main input device (i.e. keyboard).
TAB (*X*)	Used only in a Print statement and moves the cursor or print position to *X*, counted from the left-hand margin.
SPC (*X*)	Used only in a print statement and places *X* spaces between the last item printed and the next.
POS (*X*)	Returns the current horizontal printing position of the screen cursor.

Assignment and Program Control Commands

LET	Command to assign a value to a variable. Optional in many versions of BASIC.
DIM A(*X*, *Y*, *Z*)	Reserves memory space for (*X* + 1) * (*Y* + 1) * (*Z* + 1) elements of array A.
DEF FN*A* (*X*)	Defines and declares the function *A*. *X* is a dummy argument that is later substituted by a real number or variable.
FOR *X* = *A* TO *B* STEP *C* NEXT *X*	Used to execute all statements between the FOR statement and the corresponding NEXT. The number of executions is determined by the start value *A*, the final value *B* and the STEP or increment value *C*. STEP size is 1 if not specified.
GOTO *XXX*	Unconditional jump to a specified statement number (*XXX*) within the program.
GOSUB *XXX*	Branch to the subroutine at line number *XXX*.
RETURN	Marks the end of a series of statements that constitute a subroutine. Control returns to the statement following the most recent GOSUB.
IF (Condition) ...... THEN (Statement)	An instruction that allows statements to be executed only when a specified condition is met.
ON *X* GOTO 200, 510, 300	Allows multiple branching, branching to the *X*th line number in the list after the GOTO statement.
ON *X* GOSUB 200, 510, 900	Multiple subroutine branching to the *X*th line number in the list after the GOSUB.
REM	Allows comments to be included in a program; ignored during execution.

String Functions

LEN (*X*$)	Returns the number of characters in string *X*$ including blanks.
LEFT $ (*X*$, *I*)	Returns left most *I* characters of string *X*$.
RIGHT $ (*X*$, *I*)	Returns right most *I* characters of string *X*$.
MID $ (*X*$, *I*, *J*)	Returns a string of *J* characters starting with the *I*th character from string *X*$.
STRS (*X*)	Converts the numeric value *X* to a string.
VAL (*X*$)	Converts string *X*$ to a numeric value.
CHRS (*X*)	Returns the ASCII character whose code is *X*.
ASC (*X*$)	Returns the ASCII code for the first character of string *X*$.

Mathematical Functions

ABS (*X*)	Determines absolute value of *X*.

ATN(X)	Determines $\tan^{-1}(X)$ where X is in radians.
COS (X)	Determines cos (X) where X is in radians.
EXP(X)	Calculates e^x.
INT (X)	Returns largest integer $\leq X$.
LOG (X)	Determines the natural logarithm of X.
RND (X)	Generates a random number using X as the seed.
SGN (X)	Returns 1 if $X > 0$ 0 if $X = 0$ -1 if $X < 0$
SIN(X)	Calculates sin(X) where X is in radians.
SQR(X)	Determines the square root of X.
TAN(X)	Calculates tan (X) where X is in radians.

Index